.

# 2D QUANTUM METAMATERIALS

## 2018 NIST WORKSHOP

# 2D QUANTUM METAMATERIALS

## 2018 NIST WORKSHOP

Editors

### Wiley P. Kirk
University of Texas - Arlington, USA; 3D Epitaxial Technologies, LLC, USA

### John N. Randall
Zyvex Labs, LLC, USA

### James H.G. Owen
Zyvex Labs, LLC, USA

**W$\xi$ World Scientific**

NEW JERSEY · LONDON · SINGAPORE · BEIJING · SHANGHAI · HONG KONG · TAIPEI · CHENNAI · TOKYO

*Published by*

World Scientific Publishing Co. Pte. Ltd.

5 Toh Tuck Link, Singapore 596224

*USA office:* 27 Warren Street, Suite 401-402, Hackensack, NJ 07601

*UK office:* 57 Shelton Street, Covent Garden, London WC2H 9HE

**British Library Cataloguing-in-Publication Data**
A catalogue record for this book is available from the British Library.

**2D QUANTUM METAMATERIALS**
**Proceedings of the 2018 NIST Workshop**

ISBN 978-981-120-605-4

For any available supplementary material, please visit
https://www.worldscientific.com/worldscibooks/10.1142/11438#t=suppl

Desk Editor: Cheryl Heng

# 2D Quantum Metamaterials
Wiley P. Kirk
Editor
John N. Randall and James H. G. Owen
Co-editors

Recent developments in key areas of science and engineering make possible new engineered structures identified as quantum metamaterials. These new structures offer unusual properties, involving fundamental concepts such as entangled quantum states, superposition, quantum coherence, analog quantum simulation, etc., which opens a new era of technological advancement. The ideas, various approaches being pursued, and the potential technological outcomes in this field of research were discussed in a recent workshop as reported herein.

This edition is arranged in two sections. Section I — General Outcome and Conclusions, summarizes the workshop's program and the conclusions and products drawn from the workshop breakout sessions and the presentation Q&A interludes. A sufficient amount of technical detail is included in this section to support these discussions. Section II — Technical Addendum, provides additional details in the form of more expanded discussions that characterize and justify the methods and approaches of the new areas of research and as deliberated on by the attendees at the workshop. This compendium was made possible by the following authors.

| | |
|---|---|
| Gabriel Aeppli | Paul Scherrer Institute, Switzerland |
| Garnett W. Bryant | National Institute of Standards and Technology |
| Cheng Chin | University of Chicago |
| Neil J. Curson | University College London |
| Vladimir Dobrosavljevic | Florida State University |
| Kaden Hazzard | Rice University |
| Erik Henriksen | Washington University (St. Louis) |
| Ehsan Khatami | San Jose State University |
| Wiley P. Kirk | University of Texas at Arlington, 3DET |
| Norbert M. Linke | University of Maryland |
| Joseph Lyding | University of Illinois at Urbana-Champaign |
| Shashank Misra | Sandia National Laboratories |
| James H. G. Owen | Zyvex Labs |
| John N. Randall | Zyvex Labs |
| Subir Sachdev | Harvard University |
| Richard Scalettar | University of California, Davis |
| Scott Schmucker | National Institute of Standards and Technology |
| Richard Silver | National Institute of Standards and Technology |
| Bhuvanesh Sundar | Rice University |
| Ingmar Swart | Utrecht University |
| Jonathan Wyrick | National Institute of Standards and Technology |
| Neil Zimmerman | National Institute of Standards and Technology |

# Contents

# Section I — General outcome and conclusions

## 1. Introduction and objective

Wiley P. Kirk[1,2]

[1]*University of Texas at Arlington,* [2]*3D Epitaxial Technologies, LLC*

### 1.1. *Genesis, mission, and objectives of the workshop*

The idea of a Workshop on 2D Quantum Metamaterials began in 2015 when John Randall and Wiley Kirk were considering methods of utilizing scanning tunneling microscopes for atomically precise lithography and in particular for applications involving atomically precise semiconductor doping. The merits of holding a workshop became more consequential after a few program managers indicated a workshop would help with long-range planning and the development of program objectives. By late 2016, a concerted effort to formalize an organizing committee was undertaken. The individuals who agreed to be members of the organizing committee were — Richard M. Silver and Neil Zimmerman at the National Institute for Standards and Technology; Shashank Misra, Clark Highstrete, and Ezra Bussmann at Sandia National Laboratories; John N. Randall, James Owen, and Joshua Ballard at Zyvex Labs, LLC; and Wiley P. Kirk at University of Texas at Arlington and 3D Epitaxial Technologies, LLC.

The mission of the workshop as outlined by the organizing committee was to explore commonalities between fabrication, theoretical prediction, and alternative approaches to tunable quantum materials as currently realized by several approaches. Specifically, the different approaches included semiconductors doped with atomic precision, optical lattices and cold-atoms, trapped ions, superconducting circuits, approaches using entities such as quantum dots, and finally approaches based on photonic systems.

The objective of the workshop was to address and answer as best possible the following set of questions:
  (i) What is meant by a 2D quantum metamaterial from both theoretical and empirical perspectives? What might result from investigating such structures?
 (ii) Can placement of dopant atoms with atomic-scale precision in semiconductors lead to analog quantum simulations (AQS)?
(iii) What other approaches lead to AQS? How do the various approaches compare?
 (iv) To what extent can the Fermi–Hubbard model be implemented and what role might it have in expanding the horizon of physics in strongly correlated systems?
  (v) Of the various technologies that can be used to make 2D quantum metamaterials, what are the advantages and disadvantages of each?

## 1.2. *Logistics, venue, attendance statistics*

The workshop was held over a two-day period, April 25–26, 2018, as a mix of invited speakers and breakout sessions. This arrangement provided overviews of the most recent experimental and theoretical developments in each area of approach along with summarization and recommendation reports from the breakout session leaders. The workshop was held at the National Institute for Standards and Technology in Gaithersburg, Maryland, USA. Of the 87 registrants, 32 professionals were from government laboratories, agencies, and industry; 55 were professors, postdocs, and students from academia, and 5 were international participants.

## 1.3. *Organization of presentations, breakout sessions, and sponsors*

Three plenary presentations anchored the workshop's agenda:
1. Designing quantum materials in silicon, atom by atom

Shashank Misra
*Sandia National Laboratories*

2. A new toolbox for quantum many-body physics

Gabe Aeppli
*Paul Scherrer Institute, Switzerland*

3. The disordered Hubbard model: from Si:P to the high-temperature superconductors

Subir Sachdev
*Harvard University*

Ten additional presentations determined the breadth and depth of the agenda:
1. Critical quantum chaos and room temperature effects in 1D arrays of P donors in silicon

Enrico Prati
*Institute for Photonics and Nanotechnologies, Milan, Italy*

2. Fabrication of atomic-precision dopant arrays in Si using STM-based hydrogen lithography

Jonathan Wyrick
*National Institute of Standards and Technology*

3. Cold Atoms — From simulation to discoveries

Cheng Chin
*University of Chicago*

4. Quantum Simulation and Lattices in Circuit QED

Alicia Kollar
*Princeton University*

5. Quantum simulations with a semiconductor quantum dot array

Sjaak van Diepen
*Delft University of Technology*

6. Tunable solid state 2D quantum materials
   Ingmar Swart
   *Utrecht University*

7. Ultracold matter for quantum simulations: achievements, challenges, & opportunities
   Kaden Hazzard
   *Rice University*

8. Quantum simulation and quantum information with trapped ions
   Norbert M. Linke
   *University of Maryland*

9. Atom-based photonics, quantum plasmonics and many-body physics
   Garnett W. Bryant
   *National Institute of Standards and Technology*

10. Moiré is different: Bi-layer graphene as a meta material
    Philip Phillips
    *University of Illinois at Urbana-Champaign*

Breakout sessions were organized around four teams, each with a moderator and notetaker to record each team's discussions and recommendations. The following individuals served as moderators and notetakers respectively: Team 1: Joseph Lyding, University of Illinois at Urbana-Champaign (moderator), Angela Hight Walker, NIST (notetaker); Team 2: Neil Curson, University College London (moderator), Scott Schmucker, NIST (notetaker); Team 3: Irma Kuljanishvili, Saint Louis University (moderator), Igor Altfeder, Ohio State University (notetaker); Team 4: Albert Davydov, NIST (moderator), Curt Richter, NIST (notetaker). Each team was assigned the goal of producing a PowerPoint slide that captured their group's discussion and recommendations. On the last day, each team presented their team's outcome to all the registrants of the workshop. Along with general discussions, each team was asked to incorporate specific discussions on questions such as:

1. Are there general impediments to using an atomically precise dopant array for analog quantum simulations?
2. Are there particular advantages to the dopant array approach?
3. Are there particular disadvantages with the dopant array approach?
4. Are there particular lessons based on area of expertise, including other experimental realizations, that are useful for considering the dopant array approach?
5. Are there specific results from other experimental realizations (for instance, certain parameter values) that cannot be achieved by the dopant array approach?
6. Are there advantages or disadvantages of the various approaches when considering employment of manufacturing steps?
7. Finally, members of the breakout sessions were asked to consider the

degree to which an atomically precise dopant array is an attractive candidate for two-dimensional quantum materials, and in particular for executing analog quantum simulations as realized in models such as Fermi–Hubbard?

## Sponsors

The workshop was supported by the following sponsors:

### National Science Foundation
Dimitri Pavlidis
  Program Director, Division of Electrical,
  Communications & Cyber Systems (ENG/ECCS),
  Electronics, Photonics and Magnetic Devices (EPMD)
Khershed Cooper
  Program Director, Division of Civil, Mechanical and Manufacturing
  Innovation (ENG/CMMI), Nanomanufacturing (NM)

### National Institute for Standards and Technology
Rick Silver and Neil Zimmerman
  U.S. Department of Commerce,
  venue host

### Sandia National Laboratory
Shashank Misra
  Albuquerque, NM

### American Vacuum Society
New York City, N. Y.

### University of Texas at Arlington
Beth Robinson
  Department of Materials Science and Engineering

### ScientaOmicron, Inc.
North American Headquarters, Denver, CO

### Zyvex Labs, LLC
  1301 N. Plano Road, Richardson, TX

### 3D Epitaxial Technologies, LLC
  999 E Arapaho Road, Richardson, TX

## 2. Current technologies under investigation

### 2.1. *Atomically precise dopants in semiconductors: P-in-Si fabrication*

Shashank Misra[1] and Jonathan Wyrick[2]

[1]*Sandia National Laboratories,* [2]*National Institute of Standards and Technology*

The ability to make atomically precise electronic structures out of phosphorus in silicon has been developed relatively recently, the result of a concatenation of three remarkable technological developments — the ability of scanning tunneling microscopes to access the atomic scale, the discovery of a resist chemistry based on hydrogen and phosphine that creates planar dopant-based structures, and the integration of these structures into microelectronic packages that take electrical signals from millimeter length scales down to specific individual atoms. To date, much of the focus has been directed at either creating atomic-scale circuit elements[1,2] or probing the charge and spin degrees of freedom of an individual dopant atom.[3,4] Given that an individual dopant in silicon behaves as a kind of artificial atom in a silicon vacuum, the workshop explored the possibility that the dopants can be arranged into finite-size arrays where Bloch bands emerge inside the bandgap of silicon from the dopant energy levels,[5] and, in certain limits, strong electronic correlations can be added. (see Fig. 1) This presents the opportunity to study strongly correlated electrons in an environment where relevant variables can be tuned through interesting ranges by fabricating a range of arrays which differ in geometry. The arrays themselves serve as a direct analog quantum simulation of strongly correlated quantum materials — an attempt to simulate interesting quantum materials by analogy using finite periodic arrays of dopants in silicon.

Atomically precise dopant placement in silicon presents a number of key advantages and disadvantages compared to other approaches to realizing analog

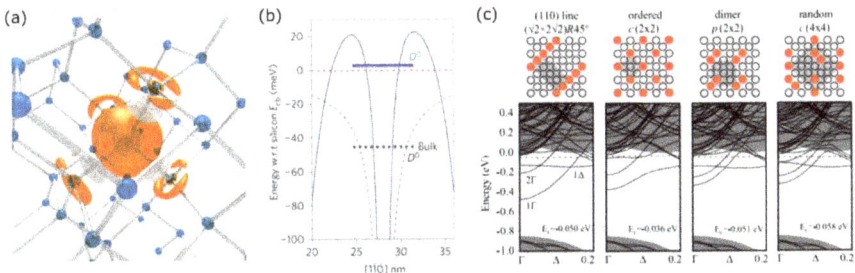

Fig. 1. From artificial atoms to Bloch bands. (a) Calculated differential charge density as a result of incorporation of a phosphorus atom in silicon. Charge density decreases in orange regions and increases in white regions. (b) Atom-like ground state energy level as simulated for a phosphorus atom in a single-atom transistor compared to the same level in bulk silicon. From Ref. [2] (c) Appearance of Bloch bands as a function of varied 2D array structures of phosphorus in silicon. From Ref. [5].

quantum simulation. Current state-of-the-art resist chemistry uses a decomposition reaction of phosphine in lithographically-defined open windows of an otherwise hydrogen-terminated silicon (100) surface. This produces a phosphorus donor within ± 1 lattice site precision 70% of the time, with no dopant resulting from the decomposition of phosphine 30% of the time. This precludes the fabrication of large arrays where every array node is a single phosphorus atom. However, given experimentally observed phosphorus incorporation statistics, 25 element arrays where individual nodes are islands of 1 to 4 donors can be fabricated with a moderately high 50% yield. Placement into a 5 × 5 array where individual 3 nm diameter islands are placed at a 5 nm pitch is expected to produce an analog simulation of a Mott–Hubbard lattice having hopping parameter $t \sim 2$ meV and Coulomb repulsion $U \sim 20$ meV. At the base temperature of a dilution refrigerator ($T \sim 10$ μeV), the antiferromagnetic ground state ($T_N \sim 200$ μeV) should be easily accessible. Outside of comparatively robust energy scales, other advantages include a well-defined and homogeneous temperature, a natural representation in terms of real electrons, and a natural means to include diatomic lattices through the development of other resist chemistries.

Implementation at the nano-scale in silicon confers its own set of advantages and disadvantages. The most straightforward implementation using the current state-of-the-art only permits measurement of electrical transport through a fabricated array; it does, however, also allow for integration with electrostatic gates that can tune the electron occupancy through a wide range (see Fig. 2), and the relatively straightforward application of large static magnetic fields. The integration of other common electronic structures may permit measurement of more than just the electronic state (metallic or insulating) of an array. Examples include single-electron transistors, which may enable a measurement of compressibility and antennas, which may enable strong coupling of electro-magnetic fields. The direct measurement of the magnetic state of a small number of spins remains an outstanding general challenge in solid-state physics but has immediate impact here via the measurement of magnetic susceptibility.

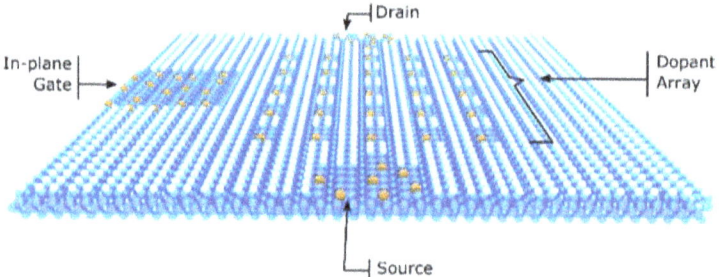

Fig. 2. Cartoon of a dopant array in silicon (pre-encapsulation) with electrostatic control via an in-plane gate and electrical transport measurement via source and drain electrodes.

An immediate opportunity in the field is to fully understand moderately sized (3 × 3, to 9 × 9) arrays. At the smaller end, model calculations can be calibrated against experimental results. Such calibration of experimental results to exactly calculable answers will be critical in developing a physical understanding of the real system. For example, it remains unclear whether the inhomogeneity from island-to-island variation in correlation energies $U$ and tunnel couplings $t$ can be ignored or would be limited. It also remains unclear the degree to which the array corresponds to a clean lattice model, or whether diagonal hopping $t'$ and extended interaction $U'$ need to be included. Developing understanding on smaller arrays opens an immediate route to having significant impact. At a larger scale, the behavior of the arrays cannot be calculated exactly; therefore, the experimental results can be used to benchmark quantum materials calculations against one another.

## Acknowledgments

Some work in this section was supported by the Laboratory Directed Research and Development program at Sandia National Laboratories, a multi-mission laboratory managed and operated by National Technology and Engineering Solutions of Sandia, LLC., a wholly owned subsidiary of Honeywell International, Inc., for the U.S. Department of Energy's National Nuclear Security Administration under contract DE-NA-0003525. This section describes objective technical results and analysis. Any subjective views or opinions that might be expressed in this section do not necessarily represent views of the U.S. Department of Energy or the United States Government.

## References

1   B. Weber, S. Mahapatra, H. Ryu, S. Lee, A. Fuhrer, T. C. G. Reusch, D. L. Thompson, W. C. T. Lee, G. Klimeck, L. C. L. Hollenberg, and M. Y. Simmons, *Ohm's Law survives to the atomic scale*, Science **335** (6064), 64–67 (2012). https://doi.org/10.1126/science.1214319.

2   M. Fuechsle, J. A. Miwa, S. Mahapatra, H. Ryu, S. Lee, O. Warschkow, L. C. L. Hollenberg, G. Klimeck, and M. Y. Simmons, *A Single-Atom transistor*, Nat. Nanotechnol. **7** (4), 242–246 (2012). https://doi.org/10.1038/nnano.2012.21.

3   H. Büch, M. Fuechsle, W. Baker, M. G. House, and M. Y. Simmons, *Quantum dot spectroscopy using a single Phosphorus donor*, Phys. Rev. B **92** (23), 235309 (2015). https://doi.org/10.1103/PhysRevB.92.235309.

4   G. C. Tettamanzi, S. J. Hile, M. G. House, M. Fuechsle, S. Rogge, and M. Y. Simmons, *Probing the quantum states of a single atom transistor at microwave frequencies*, ACS Nano **11** (3), 2444–2451 (2017). https://doi.org/10.1021/acsnano.6b06362.

8

5   D. J. Carter, N. A. Marks, O. Warschkow, and D. R. McKenzie, *Phosphorus δ-Doped Silicon: Mixed-Atom Pseudopotentials and Dopant Disorder Effects*, Nanotechnology **22** (6), 065701 (2011). https://doi.org/10.1088/0957-4484/22/6/065701.

## 2.2. Cold atoms in optical lattices

Kaden Hazzard and Bhuvanesh Sundar

*Rice University*

Ultracold matter has emerged as a versatile platform for engineering and exploring quantum phenomena. These gases of atoms or molecules have a density much lower than air, and temperatures at the nanokelvin-scale. Because the temperature is so low, quantum effects and interactions can play a crucial, even dominant, role despite the diluteness of the systems.

Ultracold matter was first created roughly three decades ago, utilizing dramatic advances in laser cooling and trapping.[1] In 1995, experiments produced Bose–Einstein condensates (BECs), realizing quantum degenerate ultracold matter and sufficiently large phase space densities to open a path to study interacting quantum matter.[2,3] Since then, ultracold experiments have furnished diverse types of strongly correlated matter,[4] pushed the frontiers of precision measurement and tests of fundamental physics,[5] and developed quantum technologies for quantum sensing[6] and quantum computation (note: Ref. [7] provides an illustrative example, summarizing the subfield of ultracold neutral atom quantum computation).

The payoff for achieving the extraordinary conditions of ultracold matter is the experimental capabilities that doing so enables. At such low temperatures, atoms' thermal motion is slowed to a crawl, and forces from light can affect the atoms. This allows experiments to leverage the extreme coherence of laser light to manipulate the atoms in novel ways. The unique tools of ultracold experiments rely in large part on this capability. Experiments routinely employ optical potentials to create traps with exquisitely controllable shapes, lattices with engineerable geometries, and disorder potentials with controllable strength. Interactions can be tuned from negligibly small to the largest allowed by quantum mechanics using Feshbach resonances.[8] Artificial gauge fields can be imposed.[9] The interaction range can be chosen to be negligibly short — effectively a delta-function $\propto \delta(r)$, the typical case in neutral ground state atoms — or long-ranged — for example, $\propto 1/r^3$ in dipolar molecules. Not only does one gain these ways to control matter, but under these conditions, the Hamiltonians that quantitatively describe these systems become standard models of condensed matter physics, such as Ising models or Hubbard models. Reviews of these capabilities in the context of quantum simulation can be found in Refs. [4,10,11].

Using these tools, experimentalists have created new forms of matter. Some of these are analogs of phases of matter that occur in solid-state systems, such as superconductors across the BEC-BCS crossover (as reviewed in Refs. [12,13]),

Mott insulators, and antiferromagnets.[14,15] More recent developments than covered in these reviews will be discussed in some detail in Section II — Technical Addendum. Some experimental realizations are entirely new forms of matter, such as time crystals[16] and many-body localized states.[17,18] Perhaps most excitingly, whole new conceptions of quantum matter have opened up, for example phases of matter that rely on long-ranged interactions or behaviors that emerge when interacting quantum systems are coherently driven far from equilibrium.[19-21]

## References

1   H. J. Metcalf and P. van der Straten, *Laser* Cooling *and Trapping*, New York: Springer-Verlager (1999).
2   C. J. Pethick and H. Smith, *Bose–Einstein condensation in dilute gases*, Cambridge: (Cambridge University Press 2008).
3   J. R. Anglin and W. Ketterle, *Bose–Einstein condensation of atomic gases*, Nature **416**, 211 (2002).
4   C. Gross and I. Bloch, *Quantum simulations with ultracold atoms in optical lattices*, Science **357**, 995 (2017).
5   M. Inguscio and L. Fallani, *Atomic Physics*, Oxford: (Oxford University Press 2013).
6   C. L. Degen, F. Reinhard and P. Cappallero, *Quantum Sensing*, Rev. Mod. Phys. **89**, 035002 (2017).
7   D. S. Weiss and M. Saffman, *Quantum computing with neutral atoms*, Physics Today **70**, 45 (2017).
8   C. Chin, R. Grimm, P. Julienne and E. Tiesinga, *Feshbach resonances in ultracold gases*, Rev. Mod. Phys. **82**, 1225 (2010).
9   N. Goldman, G. Juzeliūnas, P. Öhberg and I. B. Spielman, *Light-induced gauge fields for ultracold atoms*, Reports on Progress in Physics **77**, 126401 (2014).
10  I. M. Georgescu, S. Ashhab and F. Nori, *Quantum simulation*, Rev. Mod. Phys. **86**, 153 (2014).
11  J. I. Cirac and P. Zoller, *Goals and opportunities in quantum simulation*, Nature Physics **8**, 264 (2012).
12  G. C. Strinati, P. Pieri, G. Roepke, P. Schuck and M. Urban, *The BCS-BEC crossover: From ultra-cold Fermi gases to nuclear systems*, Physics Reports **738**, 1 (2018).
13  W. Zwerger, *The BCS-BEC Crossover and the Unitary Fermi Gas*, Heidelberg: Springer-Verlager (2012).
14  T. Esslinger, *Fermi–Hubbard Physics with Atoms in an Optical Lattice*, Annual Review of Condensed Matter Physics **1**, 129 (2010).
15  R. G. Hulet, P. M. Duarte, R. A. Hart and T.-L. Yang, *Antiferromagnetism with ultracold atoms*, in Laser Spectroscopy, World Scientific, **43**, (2016).

16  K. Sacha and J. Zakrzewski, *Time crystals: a review*, Reports on Progress in Physics **81**, 016401 (2018).

17  R. Nandkishore and D. Huse, *Many-body localization and thermalization in quantum statistical mechanics*, Annual Review of Condensed Matter Physics **6**, 15 (2015).

18  S. A. Parameswaran and R. Vasseur, *Many-body localization, symmetry and topology*, Reports on Progress in Physics **81**, 082501 (2018).

19  A. Polkovnikov, K. Sengupta, A. Silva and M. Vengalattore, *Colloquium: nonequilibrium dynamics of closed interacting quantum systems*, Rev. Mod. Phys. **83**, 863 (2011).

20  T. Langen, R. Geiger and J. Schmiedmayer, *Ultracold atoms out of equilibrium*, Annual Review of Condensed Matter Physics **6**, 201 (2015).

21  A. Lamacraft and J. Moore, - *Potential insights into nonequilibrium behavior from atomic physics*, in Ultracold Bosonic and Fermionic Gases, Chapter **7**, 177 (2012).

## 2.3. *Optical control of atomic quantum gas*

Cheng Chin

*University of Chicago*

Optical control is one of the most actively studied and promising approaches to manipulate atoms in a quantum gas. It can be applied to atoms in Bose–Einstein condensates, degenerate Fermi gases, as well as those in optical lattices. Optical control of atoms is conceptually similar to the preparation of atomic dopant arrays: individual atoms can be localized and rearranged in a sample based on strongly focused laser beams, also called "optical tweezers".[1] Optical tweezers have found wide applications in atomic and molecular physics,[2] as well as in chemistry and biology.[3] The 2018 Nobel Prize is given to Dr. Arthur Ashkin, who realized the optical tweezer in 1986.

To realize coherent control of atoms in a many-body system, quantum gases are an ideal platform. Quantum gases are intrinsically free from impurities, and the ultralow temperatures at nano Kelvin promise low thermal fluctuations and negligible dissipation. Furthermore, lower radiation power is required to control cold atoms, which is essential to suppress decoherence in generic quantum operations. Indeed, the most precise atomic clocks are based on ultracold atoms stored in an optical lattice.[4]

For cold atoms, laser beams offer high flexibility to shape the atomic potentials. One-, two-, and three-dimensional lattices can be readily formed by interfering a number of laser beams.[5] Beyond control of potential energy, interactions between atoms can also be optically or magnetically tuned based on Feshbach resonances.[6] Near a resonance, strong repulsive or attractive inter-actions between atoms can induced to enable quantum simulation of novel quantum phenomena; examples include BEC (Bose–Einstein condensate)-BCS (Bardeen–Cooper–Schrieffer superfluid) crossover,[7-9] superfluid-Mott insula-tor[10] and Efimov trimer states,[11] which was first studied in nuclear physics.

More powerful tools have been developed recently to control atoms with lasers in ways that are difficult to reproduce in electronic counterparts. These include in situ imaging of many-body systems,[12,13] synthetic gauge fields created by Raman transitions,[14] tweezer control of individual atoms,[15] modulated optical potential spatial modulation of optical potential,[16] and optical control of Feshbach resonances.[17] In particular, tuning of atomic interactions in a stable quantum gas is now realized with high temporal (nano-second) and spatial resolution (< 1 micron).[17] Finally, super resolution imaging of atomic wavefunction at nm scale has also been recently achieved.[18]

An intriguing prospect in the research of many-body physics is now within reach: the realization of an ideal quantum many-body platform with complete control of its many-body Hamiltonian. By complete control we mean its geometry, dimensionality, rearrangement of atoms, as well as their interactions with neighbors. Such a "dream quantum machine" will constitute an extremely powerful quantum simulator and can be viewed as an analog version of a universal quantum computer. Recently, independent control of 51 atoms was realized in a 1D optical lattice.[19] Future prospects for generalizing the control of individual atoms in two- and three- dimensional lattices are promising.

## References

1   A. Ashkin, *Optical trapping and manipulation of neutral particles using lasers*, PNAS **94**, 4853 (1997).

2   D. Frese, B. Ueberholz, S. Kuhr, W. Alt, D. Schrader, V. Gomer, and D. Meschede, *Single atoms in an optical dipole trap: towards a deterministic source of cold atoms*, Phys. Rev. Lett. **85**, 3777 (2000).

3   D. Grier, *A revolution in optical manipulation*, Nature **424**, 810 (2003).

4   S. L Campbell, R. B. Hutson, G. E. Marti, A. Goban, N. Darkwah Oppong, R. L. McNally, L. Sonderhouse, J. M. Robinson, W. Zhang, B. J. Bloom, and J. Ye, *A Fermi-degenerate three-dimensional optical lattice clock*, Science **358**, 6359 (2017).

5   I. Bloch, *Ultracold quantum gases in optical lattices*, Nature Physics **1**, 23 (2005).

6   C. Chin, R Grimm, P Julienne, and E Tiesinga, *Feshbach resonances in ultracold gases,* Rev. Mod. Phys. **82**, 1225 (2010).

7   C. A. Regal, M. Greiner, and D. S. Jin, *Observation of resonance condensation of fermionic atom pairs*, Phys. Rev. Lett. **92**, 040403, (2004).

8   C. Chin, M. Bartenstein, A. Altmeyer, S. Riedl, S. Jochim, J. H. Denschlag, and R. Grimm, *Observation of the pairing gap in a strongly interacting Fermi gas*, Science **305**, 1128 (2004).

9   M. W. Zwierlein, C. A. Stan, C. H. Schunck, S. M. F. Raupach, A. J. Kerman, and W. Ketterle, *Condensation of pairs of fermionic atoms near a Feshbach resonance*, Phys. Rev. Lett. **92**, 120403, (2004).

10  M. Greiner, O. Mandel, T. Esslinger, T. W. Hänsch, and I. Bloch, *Quantum phase transition from a superfluid to a Mott insulator in a gas of ultracold atoms*, Nature **415**, 6867 (2002).

11  T. Kraemer, M. Mark, P. Waldburger, J. G. Danzl, C. Chin, B. Engeser, A. D. Lange, K. Pilch, A. Jaakkola, H. C. Nägerl, and R. Grimm, *Evidence for Efimov quantum states in an ultracold gas of caesium atoms*, Nature **440**, 7082 (2006).

12  N. Gemelke, X. Zhang, C. L. Hung, and C. Chin, *In situ observation of incompressible mott-insulating domains in ultracold atomic gases*, Nature **460**, 995 (2009).

13  W. S. Bakr, J. I. Gillen, A. Peng, S. Fölling, and M. Greiner, *A quantum gas microscope for detecting single atoms in a Hubbard-regime optical lattice*, Nature **462**, 74 (2009).

14  Y.-J. Lin, R. L. Compton, K. Jiménez-García, J. V. Porto, and I. B. Spielman, *Synthetic magnetic fields for ultracold neutral atoms*, Nature **462**, 628 (2009).

15  A. M. Kaufman, B. J. Lester, and C. A. Regal, *Cooling a single atom in an optical tweezer to its quantum ground state*, Phys. Rev. X **2**, 041014 (2012).

16  André Eckardt, *Colloquium: Atomic quantum gases in periodically driven optical lattices,* Rev. Mod. Phys. **89**, 011004 (2017).

17  Logan W. Clark, Li-Chung Ha, Chen-Yu Xu, and Cheng Chin, *Quantum dynamics with spatiotemporal control of interactions in a stable Bose–Einstein condensate,* Phys. Rev. Lett. **115**, 155301 (2015).

18  M. McDonald, J. Trisnadi, K. Yao, and C. Chin, *Super-resolution microscopy of cold atoms in an optical lattice*, ArXiv:1807.02906

19  Hannes Bernien, Sylvain Schwartz, Alexander Keesling, Harry Levine, Ahmed Omran, Hannes Pichler, Soonwon Choi, Alexander S. Zibrov, Manuel Endres, Markus Greiner, Vladan Vuletić, and Mikhail D. Lukin, *Probing many-body dynamics on a 51-atom quantum simulator,* Nature **551**, 579–584 (2017).

## 2.4. *Trapped ion technologies*

### Norbert Linke

*University of Maryland*

Control over individually trapped ions has developed rapidly in the past two decades. Thanks to their unmatched coherence properties,[1] and long-range coupling via the Coulomb interaction,[2] trapped ions have become a versatile platform employed for a large variety of experimental goals. The steady stream of ground-breaking results range from quantum simulation in the realm of quantum many-body phenomena,[3,4] to quantum chemistry studies (including the control of all degrees of freedom in molecules),[5] to some of the most precise atomic clocks ever built.[6] Trapped ions are also one of the leading contenders for realizing a universal circuit model quantum computer.[7,8]

A good candidate system to create a quantum computer or quantum simulator has to fulfil a set of requirements such as the ability to perform state initialization, a set of operations, and read-out.[9,10] Additionally, in order to be useful, the quantum system needs to have low or well controlled coupling to the environment in order to exhibit coherence times that are long compared to the evolution under investigation. Furthermore, we need to be able to control sufficient degrees of freedom to usefully model a quantum problem, i.e., in a quantum computer this corresponds to being able to scale to a sufficiently large number of qubits.

A single atom trapped in isolation from any other atoms or molecules is a pristine quantum system under this paradigm. Atoms have energy levels with transitions in the visible or ultraviolet range, which can be addressed with laser light. Due to these energy scales, experiments can be conducted in room temperature apparatus. Information about their state can be gathered using state-dependent fluorescence. Atoms are used to create highly accurate frequency references since they form identical standard systems. So, scaling of the quantum device can be achieved by adding more atoms, each of which will have identical properties. To achieve a Hamiltonian of interest, we need to engineer strong interactions in the system which is particularly advantageous to do with ions; since, as charged atoms, they interact strongly via the Coulomb force. A set of ions trapped in a common confining potential will exhibit a common set of motional modes. By coupling these modes to the internal (spin) degrees of freedom, we can engineer interactions between the spins, essentially using the motional modes as an information carrier or bus.[11] Additional details can be found in Section II — Technical Addendum of this document.

# References

1   T. P. Harty, D. T. C. Allcock, C. J. Ballance, L. Guidoni, H. A. Janacek, N. M. Linke, D. N. Stacey, and D. M. Lucas, *High-fidelity preparation, gates, memory, and readout of a trapped-ion quantum bit*, Phys. Rev. Lett. **113**, 220501, Nov (2014).

2   K. Kim, M.-S. Chang, R. Islam, S. Korenblit, L.-M. Duan, and C. Monroe, *Entanglement and tunable spin-spin couplings between trapped ions using multiple transverse modes*, Phys. Rev. Lett. **103**, 120502, Sep (2009).

3   C. Schneider, D. Porras, and T. Schaetz, *Experimental quantum simulations of many-body physics with trapped ions*, Rep. Prog. Phys. **75**, 2, 024401 (2012).

4   R. Islam, C. Senko, W. C. Campbell, S. Korenbilt, J. Smith, A. Lee, E. E. Edwards, C.-C. J. Wang, J. K. Freericks, and C. Monroe, *Emergence and frustration of magnetism with variable-range interactions in a quantum simulator*, Science **340**, 583–587 May (2013).

5   C.-W. Chou, C. Kurz, D. B. Hume, P. N. Plessow, D. R. Leibrandt, and D. Leibfried, *Preparation and coherent manipulation of pure quantum states of a single molecular ion*, Nature **545**, 203–207 May (2017).

6   N. Huntemann, C. Sanner, B. Lipphardt, C. Tamm, and E. Peik, *Single ion atomic clock with $3 \times 10^{-18}$ systematic uncertainty*, Phys. Rev. Lett. **116**, 063001 Feb (2016).

7   S. Debnath, N. M. Linke, C. Figgatt, K. A. Landsman, K. Wright, and C. Monroe, *Demonstration of a small programmable quantum computer module using atomic qubits*, Nature **536**, 63–66 Aug (2016).

8   T. Monz, D. Nigg, E. A. Martinez, M. F. Brandl, P. Schindler, R. Rines, S. X. Wang, I. L. Chuang, and R. Blatt, *Realization of a scalable Shor algorithm*, Science **351**, 6277, 1068–1070 (2016).

9   D. P. DiVincenzo, *The physical implementation of quantum computation*, Fortschritte der Physik **48**, 911, 771–783 (2000).

10   C. A. P´erez-Delgado and P. Kok, *Quantum computers: Definition and implementations*, Phys. Rev. A **83**, 012303 Jan (2011).

11   D. J. Griths, *Introduction to Electrodynamics*. Prentice Hall, 3rd Ed. (1999).

## 2.5. *Macroscale circuits via superconductors*

Wiley P. Kirk[1,2]

[1]*University of Texas at Arlington,* [2]*3D Epitaxial Technologies*

Dr. Alicia Kollar, a postdoc in Prof. Andrew Houck's group at Princeton, provided the workshop an up to date review about how superconductors are used in quantum entanglement applications and specifically for the formation of qubits in quantum computation. The current status of superconducting transmon qubits and their use in quantum computation applications was recently reviewed by Oliver *et al.*[1] However, general discussions in the workshop breakout sessions concluded that problems associated with scaling up and implementing quantum metamaterials concepts via superconducting macroscale circuit approaches did not have a particularly promising future. Although superconductors are widely acknowledged to be capable of demonstrating quantum coherence effects on a macroscopic level, their future technological and scientific role is probably more akin to how vacuum tube science and technology has played out against solid-state semiconductor science and technology.

## References

1 P. Krantz, M. Kjaergaard, F. Yan, T. P. Orlando, S. Gustavsson, and W. D. Oliver, *A Guide to Superconducting Qubits for Quantum Information Engineers*, arXiv:1904.06560.

## 2.6. *Quantum simulation using quantum dot arrays*

Neil Zimmerman

*National Institute of Standards and Technology*

A variety of predicted and measured results in electrostatically-confined semiconducting quantum dot arrays have been pursued in recent years. The attraction of pursuing analog quantum simulations (AQS) in semiconductor quantum dot arrays is driven by the exciting possibility of getting into the parameter regime $U = k_B T$. [Additional details about the interaction energy $U$ can be found below in sections 4.1, 4.2.1, 4.4, and in Section II — Technical Addendum.] It is clear from what follows that theoretical proposals involving semiconducting quantum dots are well in advance of experimental achievements in this area of research. While there have been some preliminary experimental results, the field is impeded in large part by the difficulty of reducing the disorder sufficiently to achieve the Hubbard model conditions. On the other hand, these devices do appear feasible for simulating the Anderson–Hubbard model, which explicitly considers disorder. Please refer to Section II — Technical Addendum for details.

Electrostatically-confined semiconducting quantum dots are made in a variety of ways, generally as surface-gated bulk crystalline semiconductors or deposited nanowires; these devices are generally controlled by gate voltages and are measured at cryogenic temperatures. Other applications of quantum dot devices include quantum coherent manipulation, accurate standards of electrical current, and conventional computing including quantum cellular automata.

There have been a variety of theoretical proposals to use an array of quantum dots for AQS, generally using surface-gated quantum dots on planar semiconductors such as GaAs or Si. These proposals focus on (i) being able to control the parameters of the Fermi–Hubbard or Heisenberg or other Hamiltonians using gate voltages, (ii) detection/read-out of charge or spin degree of freedom, and (iii) predictions of the parameter regimes achievable.

On the experimental side, there has been substantial progress in "arrays" with a maximum of three or four quantum dots; the simulated results do not extend beyond what can be achieved by theory or classical analogue simulations. In addition, for larger arrays, there have been some preliminary results; in general, the issue of disorder is substantially impeding progress.

## 2.7. *Other approaches for 2D quantum metamaterials*

Ingmar Swart

*Utrecht University*

Quantum matter offers a plethora of high-quality materials that exhibit exotic physical characteristics and phenomena such as quantized currents at edges of otherwise insulating materials, chiral magnetism, topology, and emergent behavior that may be used for information and sensing technology. Such quantum materials exhibit remarkable transport phenomena, not only based on manipulating the electronic but also the spin, valley, or orbital degrees of freedom, creating much more versatility and functionality compared to traditional materials. However, the leap beyond serendipity toward predictability relies on developing a fundamental description of the building blocks that lead to such quantum states of matter, fed by rational experimental design.

Low-temperature scanning tunneling microscopy (STM) has become a powerful tool to both create and characterize quantum matter at the atomic scale. Thus far, artificial 1D and 2D lattices of electrons and spins have been realized using a variety of material platforms. These include adatoms and molecules on surfaces,[1-4] vacancies in insulating layers,[5-7] and 2D electron gasses.[8-10] Manual assembly of lattices atom-by-atom can be time consuming. Software tools that allow the autonomous assembly of structures will greatly enhance the number of structures that can be investigated, as well as the size of these structures.[9,11]

The main driver for this work is that designer lattices, built atom-by-atom, in the STM offer a way to understand the fundamental behavior of fascinating states of matter, such as quantum spin liquids, topological superconductors, Dirac quasiparticles, or the dynamics of quantum spin lattices. Therefore, creating designer lattices offers a unique and synergistic approach to understand and discover new states of matter. The main disadvantage is that the as-built structures are only stable at low temperature. The temperature depends on the material system and can be as high as 77 K.[5]

## References

1  C. F. Hirjibehedin, C. P. Lutz and A. J. Heinrich, *Spin coupling in engineered atomic structures,* Science **312**, 5776, 1021–1024 (2006).

2  D. Wegner, R. Yamachika, X. W. Zhang, Y. Y. Wang, T. Baruah, M. R. Pederson, B. M. Bartlett, J. R. Long and M. F. Crommie, *Tuning molecule-mediated spin coupling in bottom-up-fabricated vanadium-tetracyanoethylene nanostructures,* Phys. Rev. Lett. **103**, 8 (2009).

3    A. A. Khajetoorians, J. Wiebe, B. Chilian and R. Wiesendanger, *Realizing all-spin–based logic operations atom by atom*, Science **332**, 6033, 1062–1064 (2011).

4    N. Nilius, T. M. Wallis and W. Ho, *Development of one-dimensional band structure in artificial gold chains*, Science **297**, 5588, 1853–1856 (2002).

5    F. E. Kalff, M. P. Rebergen, E. Fahrenfort, J. Girovsky, R. Toskovic, J. L. Lado, J. Fernández-Rossier and A. F. Otte, *A kilobyte rewritable atomic memory*, Nature Nanotechnology **11**, 926 (2016).

6    J. Girovsky, J. L. Lado, F. E. Kalff, E. Fahrenfort, L. J. J. M. Peters, J. Fernández-Rossier, A. F. Otte, *Emergence of quasiparticle Bloch states in artificial crystals crafted atom-by-atom*, Sci. Post Phys. **2**, 020 (2017).

7    R. Drost, T. Ojanen, A. Harju and P. Liljeroth, *Topological states in engineered atomic lattices*, Nature Physics **13**, 668 (2017).

8    K. K. Gomes, W. Mar, W. Ko, F. Guinea, and H. C. Manoharan, *Designer Dirac fermions and topological phases in molecular graphene*, Nature **483**, 7389, 306–310 (2012).

9    M. R. Slot, T. S. Gardenier, P. H. Jacobse, G. C. P. van Miert, S. N. Kempkes, S. J. M. Zevenhuizen, C. Morais Smith, D. Vanmaekelbergh and I. Swart, *Experimental realization and characterization of an electronic Lieb lattice*, Nature Physics **13**, 672 (2017).

10   L. C. Collins, T. G. Witte, R. Silverman, D. B. Green and K. K. Gomes, *Imaging quasiperiodic electronic states in a synthetic Penrose tiling*, Nature Communications **8**, 15961 (2017).

11   R. J. Celotta, S. B. Balakirsky, A. P. Fein, F. M. Hess, G. M. Rutter, and J. A. Stroscio, *Autonomous assembly of atomically perfect nanostructures using a scanning tunneling microscope*, Review of Scientific Instruments **85**, 121301 (2014).

## 2.8. *Photon systems*

Garnett Bryant

*National Institute of Standards and Technology*

There are several ways in which photons (photonics) could play an important role in the development of atom-based planar (1D or 2D) materials. The focus of this workshop has been on the use of arrays of dopants or patches of dopants to perform quantum simulations, especially of the Fermi–Hubbard model. Such proposals typically envision some form of exchange coupling between spin qubits on adjacent donors to mediate the interactions and hopping in a Fermi–Hubbard model. However, such interactions can be very sensitive to inter-donor spacing with order of magnitude variations for changes in dopant position of one lattice site. Through spin-photon conversion using optical transitions in deep donors like $Se^+$ in Si, photons could be used, instead of exchange, to provide coupling between spin qubits. More details are provided in the Section II — Technical Addendum. Such a capability opens another way to operate arrays of dopants for simulations.

Dopant arrays could also be used for a variety of compelling photonic applications. Several possibilities were discussed in the workshop to highlight what might be possible. For example, it has recently been shown theoretically that a single 2D layer of a transition metal dichalcogenide can act as a perfect mirror. Within the last few months, large excitonic reflectivity by monolayers of $MoSe_2$ has been observed experimentally. These results suggest the possibility of constructing entire optical circuits using single layers of atomic based structures if dopants with optical transitions are used or if the arrays can support excitonic states.

Other photonic functionalities could be engineered with dopants in Si. Quantum memories for photonic qubits have been long sought. Rare-earth dopants in Si could provide micrometer-scale photonic memories when a cavity is used to provide enhanced coupling and the photon states are stored by slowing or stopping the light with transitions by the dopants. These quantum memories have been implemented with randomly placed rare-earth dopants. 2D arrays based on STM lithography could provide even greater functionality and open ways to integrate quantum memories with a diverse set of quantum technologies. More details are given in the Section II — Technical Addendum. Atom-based clocks have become the standard for time-keeping, providing the precision needed for modern commerce and banking, precision tracking and mapping, and high-precision frequency standards. On-chip atomic clocks could be implemented with dopants in Si provided the dopants have transitions insensitive to environmental effects. Such clock transitions have been demonstrated for Bi in Si. Again, the use of 2D dopant arrays based on STM lithography might provide greater functionality and more ways to integrate atomic clocks with a diverse range of quantum technologies and for a wide range of applications that need on-chip clocks, for example, to do tracking in local environments such as a building or warehouse without relying on GPS, or to stabilize or identify signals in noisy environments.

Single-photon emission from individual, localized defects in atomically thin semiconductor layers, such as $WSe_2$ and $WS_2$, have been demonstrated recently. Such arrays of quantum emitters in two-dimensional materials offer promising applications in quantum computing, quantum communication, quantum sensing and fundamental quantum science. Atom-based arrays in Si could provide another way to achieve ordered arrays of quantum emitters in planar structures, allowing for atomically precise placement of the emitters and controlled integration with the other atom-based optical elements already mentioned, such as clocks, memories, and routing elements to create quantum optical circuits.

So far, we have discussed ways in which photons and photonics could play an important role in the development or application of atom-based planar (1D or 2D) materials. The focus of the workshop was on the use of arrays of dopants or patches of dopants to perform quantum simulations, especially of the Hubbard model. Typically, such simulations are done to determine ground state phase diagrams and understand the competing effects of many-body interactions, single-particle hopping and quantum fluctuations in simple models. Applications of dopant arrays to photonics require an understanding of many-body excited states and response. Dopant arrays could be used both to do these excited state simulations and to realize the atom-based photonic structures being simulated. Dopant arrays could be used both to simulate excited state dynamics, many-body entanglement, collision, and interference of many-body excitations and to implement atom-scale quantum and photonic applications. Quantum simulations to find excitation spectra and many-body states can be done computationally for small systems up to about 30 atoms for a 1D chain of atoms with spinless electrons. Details are provided in the Section II — Technical Addendum. Direct diagonalization of even a few states becomes intractable for longer chains. Including additional effects, such as spin or coupling to a parallel chain of atoms to model two dimensional effects or treating larger systems will require quantum simulations with dopant arrays.

# 3. Definition of atomically precise fabrication and quantum metamaterials

Neil Curson

*University College London*

Technological progress in the pursuit of fabricating ever smaller structures has been relentless over the last few decades, spawning the field of nanotechnology in the process. Nanotechnology, commonly defined as the manipulation of matter with at least one dimension sized from 1 to 100 nanometers, has revolutionized numerous areas of science and technology. A major technological driver behind the reduction of feature sizes to the nanoscale regime has been the continuous improvement in the nanofabrication techniques themselves. However, a new technological dawn is already emerging, *viz.* ones based directly on the laws of quantum mechanics.[1] For example, quantum technologies are already being commercialized by companies such as D-Wave, M Squared and MagiQ.[2-4] To meet the fabrication challenges associated with exploiting quantum effects, fabrication techniques have to be pushed to their ultimate limits. In the solid state, this leads us to atomically precise fabrication,[5,6] where single atoms become the building blocks of the structures being made. Hence the ability to routinely locate, image and manipulate single atoms, and structures consisting of single atoms, is the next great fabrication challenge.

One application that has evolved hand-in-hand with evolving fabrication technology is that of metamaterials. Historically, a metamaterial is defined as an engineered periodic structure (typically with a pitch of ~ 100–1000 nm) that is not found in nature and is designed to alter the properties of the electromagnetic radiation with which it interacts.[7] For example, metamaterials have been fabricated that exhibit optical cloaking,[8] negative refractive index,[9] and super-lensing effects.[10] Other types of metamaterials such as mechanical metamaterials also exist.[11] The quantum version of a metamaterial is one whose periodicity is small enough such that quantum effects become dominant. Phenomena could be explored such as Mott physics with independently controllable dopant disorder and density,[12] electron lensing based on doping modulation and improved quantum phase memory devices. In the ultimate limit of making metamaterials from individual atoms, 2D periodic structures are particularly appealing as they can be conveniently embedded in the lattice of the surface of a material. Thus, atomic scale confinement perpendicular to the surface comes for free.

**References**

1   A. Acín, I. Bloch, H. Buhrman, T. Calarco, C. Eichler, J. Eisert, D. Esteve, N. Gisin, S. J. Glaser, F. Jelezko, S. Kuhr, M. Lewenstein, M. F. Riedel, P. O. Schmidt, R. Thew, A. Wallraff, I. Walmsley, and F. K. Wilhelm, *The quantum technologies roadmap: a European community view*, New J. Phys. **20**, (2018). 080201 DOI: 10.1088/1367-2630/aad1ea

2   https://www.dwavesys.com/home

3   http://www.m2lasers.com/quantum.html

4   http://www.magiqtech.com

5   S. R. Schofield, N. J. Curson, M. Y. Simmons, F. J. Rueß, T. Hallam, L. Oberbeck, and R. G. Clark, *Atomically precise placement of single dopants in Si*, Phys. Rev. Lett. **91**, 136104 (2003). DOI: 10.1103/PhysRevLett. **91**, 136104

6   S. Fölsch, J. Martínez-Blanco, J. Yang, K. Kanisawa, and S. C. Erwin, *Quantum dots with single-atom precision*, Nature Nanotechnology **9**, 505 (2014). DOI: 10.1038/NNANO. **129**, (2014)

7   S. Zhu and X. Zhang, *Metamaterials: artificial materials beyond nature*, National Science Review **5**, 131 (2018). DOI: 10.1093/nsr/nwy026

8   D. Schurig, J. J. Mock, B. J. Justice, S. A. Cummer, J. B. Pendry, A. F. Starr, D. R. Smith, *Metamaterial electromagnetic cloak at microwave frequencies*, Science **314**, 977 (2006). DOI: 10.1126/science.1133628

9   J. Valentine, S. Zhang, T. Zentgraf, E. Ulin-Avila, D. A. Genov, G. Bartal, and X. Zhang, *Three-dimensional optical metamaterial with a negative refractive index*, Nature **455**, 376 (2008). DOI: 10.1038/nature07247

10  N. Fang, H. Lee, C. Sun, X. Zhang, *Sub–Diffraction-Limited Optical Imaging with a Silver Superlens*, Science **308**, 534 (2005). DOI: 10.1126/science.1108759

11  K. Bertoldi, V. Vitelli, J. Christensen, and M. van Hecke, *Flexible mechanical metamaterials*, Nature Reviews Materials **2**, 17066 (2017). DOI:10.1038/natrevmats.66.(2017)

12  E. Prati, M. Hori, F. Guagliardo, G. Ferrari and T. Shinada, *Anderson–Mott transition in arrays of a few dopant atoms in a silicon transistor*, Nature Nanotechnology **7**, 443 (2012). DOI: 10.1038/NNANO.2012.94

### 3.1. *Fabrication challenges to overcome — precision, scale*

Joseph Lyding

*University of Illinois at Urbana-Champaign*

For atomically precise fabrication, scanned probe microscopy (SPM), particularly scanned tunneling microscopy (STM), is the tool of choice. This immediately brings to the forefront several challenges that stand between experimental demonstrations and a technology platform. One of these, of course, is the slow throughput SPM-based systems. This issue is addressed in Sec. I (3.3) of this report. Other challenges, pertaining to the scanning units, are piezoelectric creep and hysteresis, and thermal drift. There are a host of issues associated with probe tips, including tip geometry control, oxide coatings and contamination, and stability and lifetime under lithography conditions.

Time-dependent piezoelectric creep is an ubiquitous problem in SPM that interferes with the positioning accuracy and stability of the probe relative to the surface. It results from residual domain wall movement in the piezoelectric material following a voltage change across the piezoelectric element. As such, compensating for creep is complex in that it depends on the magnitude and history of piezo voltage changes. Piezoelectric hysteresis is a similar problem that generally requires imaging with unidirectional scan traces. Barring developments in creep and hysteresis-free materials, software correction that applies a time and history-dependent additive piezo voltages, is a viable approach to null creep and hysteresis. An example of this is the ZyVector system developed by Zyvex Labs.[1]

Thermal drift can be minimized by cryogenic operation, stable temperature control at high temperatures, minimization of the tip-sample mechanical loop, use of low thermal expansion coefficient materials, and by combinations of these factors in a thermally compensated scanner design.[2]

MEMS-based scanners offer the potential to address many SPM issues. Likewise, the small size of a MEMS scanner minimizes thermal drift issues as well. One complication, though, is the integration of probe tips with MEMS scanners. However, MEMS-based scanners may become sufficiently low-cost that it is economical to replace the entire scanner when its tip fails.

Probes issues prevent SPM and STM-based lithography from having the day-to-day turnkey consistent performance of traditional scanned beam instruments. While the probe apex must be sharp on the nanometer to atomic scale, the general probe shape is also important. Oxide or other contamination layers on probes interfere with lithography performance and can lead to changes in performance during writing. While tungsten, and platinum-iridium probes have been the mainstays for STM lithography, most groups produce their own probes. Probes of repeatable geometry and functionality are not yet commercially available. However, newer sharpening techniques like field-directed sputter sharpening[3] are showing promise for better geometry control and an expanding pallet of probe materials, including probes with super hard coatings.

The scaling issue is discussed in Sec. I (3.3) below with regard to having arrays of MEMS scanners. One implementation of this could involve simplified MEMS devices that only provide one axis of motion along the tip-sample $z$-direction. One can envision an array of these scanners placed on a precision $(x,y,z)$ stage such as the ones currently used in photolithography steppers. The entire array would be scanned together with nanometer scale resolution and overlay accuracy, while the actual patterning would be defined by the individual $z$-axis and voltage control of the writing tips. Small advances in interferometric stage control would move the resolution and overlay to atomic precision.

**References**

1 https://www.zyvexlabs.com/apm/products/zyvector/
2 J. Lyding, S. Skala, J. Hubacek, R. Brockenbrough, and G. Gammie, *Variable-temperature scanning tunneling microscope*, Rev. Sci. Instrum. **59** (9), 1897–1902 (1988).
3 S. Schmucker, N. Kumar, J. Abelson, S. Daly, G. Girolami, M. Bischof, D. Jaeger, R. Reidy, B. Gorman, J. Alexander, J. Ballard, J. Randall, and J. Lyding, *Field-directed sputter sharpening for tailored probe materials and atomic-scale lithography*, Nat. Commun., **3** (2012).

## 3.2. *Advantages and disadvantages from a state-of-the-art perspective*

Richard Silver and Scott Schmucker

*National Institute of Standards and Technology*

There are many sizable advantages to a quantum metamaterial platform in which dopant atoms are placed within a solid-state architecture in Si. The use of Si as a medium provides compatibility with well-established semiconductor processing techniques while also leveraging more than 15 years of research in pursuit of STM-based Si:P fabrication. Additionally, this platform provides a mechanism for transitioning the foundational 2D cold atom array into a solid-state environment. Although fabrication of these materials requires a clean UHV environment, once encapsulated, materials are long-lived and air-stable. However, an STM-based fabrication process carries its own set of challenges. In addition to the inherently serial nature of tip-based nanofabrication, atomic-scale STM patterning demands stable, atomically sharp metal probes and reliable procedures for preparing atomically-flat and clean Si(100) surfaces. Future development of layered three-dimensional devices will require reduced-temperature Si(100) surface preparation, further complicating sample preparation.

Beyond STM lithography, the fabrication of clean and functional 2D quantum metamaterials in Si requires a breadth of external capabilities that can limit the ability of a single research laboratory to undertake efforts in the field, likely demanding collaborative efforts between labs and institutions. During fabrication, extremely clean vacuum conditions are necessary to preserve clean surfaces while devices and large-area contact pads are written and to minimize contamination in the Si epitaxial layer. Traditionally, UHV pressures at or below $1 \times 10^{-10}$ Torr are utilized for this purpose; however, present efforts seek to improve these standard operating conditions to $\approx 10^{-12}$ Torr, with plans to operate in extreme high vacuum (XHV) conditions in the future. Once devices are fabricated *in vacuo*, further processing requires cleanroom infrastructure with electron-beam lithography, metal deposition, and rapid thermal annealing. Measurements must then be performed at a variety of cryogenic temperatures: most measurements require temperature below $\approx 10$ K to freeze charge carriers in the Si substrate with many measurements demanding millikelvin temperatures and the use of a dilution refrigerator and high magnetic fields. Finally, the process of designing and analyzing devices and metamaterials requires a modeling effort operated in parallel, which can provide appropriate geometrical bounds on device designs while also validating the process. Significantly, each of these processes is incorporated into an ongoing feedback loop: information must be continually exchanged between STM lithographers, process engineers, low-temperature measurement experts, and solid-state materials and device modelers.

Also, another critical infrastructure concern is safety. Dopant atoms are typically introduced by dissociative chemisorption of highly toxic gas species (e.g., phosphine, diborane, *etc.*) on the patterned Si(100) surface. Adding this capability to any laboratory requires appropriate risk assessment and the incorporation of controls for risk management. An incomplete subset of such controls is described. Each laboratory must develop its own process for risk management, which provides a breadth of controls under which multiple simultaneous failures are required before hazardous exposure can occur. Where possible, sub-atmospheric gas cylinders are selected, which require toxic-gas pumping equipment to extract the hazardous species. Consequently, all gas lines are held below atmospheric pressure. Where possible, all gas lines containing hazardous species are contained fully within a vented gas cabinet with appropriate gas detection and alarm systems. When gas lines must extend beyond the cabinet, either the pressure is first reduced to $\leq 1 \times 10^{-3}$ Torr or the gas is transported through welded double-walled tubing with an evacuated and sensor-monitored outer shell.

### 3.3. *Prospects for nanomanufacturing and the scale up problem*

John N. Randall

*Zyvex Labs*

Hydrogen Depassivation Lithography (HDL) using a Scanning Tunneling Microscope (STM) is the current means of creating 2D arrays of dopant atoms in a buried Si (100) plane.[1] The HDL is used to create patterns of clean Si dimers on a Si (100) 2×1 H passivated Si. The clean Si is very reactive and $PH_3$ will stick to the clean surface and does not stick to a H passivated surface. A short anneal to incorporate the donor dopant P atoms followed by low temperature Si epitaxy leaves, in the area that was patterned, roughly ¼ of the atoms (now buried by a few 10's of nm of Si epitaxy) as activated donors. The same patterning, incorporation and epitaxial growth processes can produce tunnel barrier contacts at the edge of the array and other electrodes a little farther away that can apply an electrostatic bias to the array. Additional conventional deposition and patterning processes can create metal contacts that can be attached to different measurement devices or circuitry. The only significant limitation to scaling up is the very slow serial HDL patterning. Everything else can be done with conventional semiconductor style nano/micro fabrication processes.

HDL is a form of electron beam lithography. It produces a very small spot of electrons that are scanned across the surface of a substrate and exposes a resist. It has sub-nm resolution and the ability to remove a single H atom from the Si surface.[2] However, it is a very different regime of e-beam lithography than conventional e-beam Lithography (CEBL). In HDL the electron energies are roughly three orders of magnitude lower; and the beam currents are at least two orders of magnitude higher, comparing the highest resolution modes for HDL and CEBL. The scanning is done mechanically, and the resist is a monolayer of H that self-develops.[3] HDL has the added advantage of producing a much sharper contrast than CEBL, which has a multi-Gaussian point-spread function where the deposited exposure energy does not reduce to 10% of the maximum until a radius of roughly 4 nm.[4] HDL on the other hand, has a double exponential in the exponential decay of current with tunneling distance and a highly nonlinear dependence of exposure efficiency on current because the exposure process is multi-electron.[5] This results in the exposure efficiency dropping by eight orders of magnitude at a radial distance of 0.5 nm.[5]

The tools that are in use today to do HDL are effectively variable-spot size, variable-energy, vector-scan instruments. The most significant difference from conventional e-beam lithography is the resist that is extremely inefficient because the exposure process for the small spot mode (2–5 V) is a multi-electron process. While there is a higher bias (8–80 V) mode, which is a much more efficient exposure mode,[6] the key to the sort of patterns that will make 2D quantum metamaterials is the low-bias, multi-electron exposure process that is very inefficient and a significant bottleneck to scaling. As of 2018 the typical exposure rate for the atomic-resolution low-bias exposure mode is 100 H atoms per second. It is much faster than pushing atoms around on surfaces such as done by Sander Otte[7] and Ingmar Swart.[8] It is also more than capable of patterning large arrays, which can be used to make large arrays of dopant atoms. Zyvex Labs have demonstrated patterning of 100×100 5 nm box arrays. The workshop concluded that smaller arrays (10×10 for instance) will likely be more useful for Fermi–Hubbard model experiments. However, at 100 atoms a second, HDL is about 1000× slower in exposure area/unit time than conventional e-beam that can pattern at 5 nm resolution — which is already painfully slow.[5] While it is conceivable that the exposure speed could be increased by one or maybe even two orders of magnitude, this likely would not be fast enough to manufacture anything but very niche market products. Although there are some very high-resolution masked patterning processes, the need to align to the crystal lattice to achieve atomic resolution makes a masked approach seem an unlikely candidate for atomic resolution patterning via HDL or for that matter any other known patterning approach. The problems with a masked HDL approach include making a stencil mask with atomic resolution in patterned openings and pattern placement, electron scattering from the stencil mask edge, penumbral blur from a finite-sized electron source, and the change in atomic position of atoms at every step edge.

In spite of the difficulties of going parallel with e-beam lithography, a massively parallel array of STM scanners is a path worth considering. We should recognize that there the two major problems of taking conventional e-beam lithography parallel: space charge effects and wiring cross talk. The space charge effects are created by a large number of high current density electron beams each with varying currents. Some of the interactions can be modelled and corrected for, but other interactions they are stochastic and cannot be corrected.[9,10] The problem gets exponentially worse as the number of beams is increased. This cross-talk problem results from a need to control with high current or voltage the blanking or deflection of a large array of electron beams. As the size of the array grows, then getting the high voltage or currents into the array become a wiring and crosstalk nightmare.

Happily, both of these problems, which have plagued the parallelization of conventional e-beam lithography, can be avoided in a massively parallel array of

STM scanners. The approach would be to replace the conventional piezoelectric tube scanner with a Micro Electro Mechanical System (MEMS) scanner. This allows the implementation of three-degree-of-freedom scanners that can be put on a footprint as small as $90 \times 110$ $\mu m^2$.[5] This would mean a density of 10,101 scanners/tips $cm^{-2}$. But with more than 10,000 tips squeezed into a square cm, won't space charge be a major problem? The answer is that it won't be a problem at all. With the "beam path" of approximately 1 nm and the tips separated by 90 $\mu m$ there will not be any appreciable space charge effect. In fact, with the low bias mode, electrons tunnel from occupied states in the tip into unoccupied states in the Si in a manner that charge build up in the intervening vacuum is circumvented. Also, the wiring and crosstalk problems can be avoided by making the MEMS, "Smart MEMS" by including a CMOS microcontroller with each MEMS scanner.[5] In this way the individual scanners could be placed on a data bus and a power bus. The scan area of each MEMS scanner would be significantly smaller than its footprint. The entire array could be placed on an X–Y global platform, so the scan field patterns could be stitched together, and the stitched patterns of each scanner could then also be stitched together. Using the data bus, each scanner of the entire addressable array receives high-level patterning instructions, carries them out and reports back to the master controller that it is ready for new instructions. We believe that this architecture would reduce the complexity of scaling to larger numbers of scanners from an exponential growth of difficulty to a sub-linear one. It is not inconceivable that the area of a 300 mm wafer could be filled with scanners resulting in parallelism in the range of 7 million MEMS X, Y, Z scanners and a tip for each scanner. We point out that DARPA has funded MEMS-based Scanning Probe Scanners and Integrated Circuit Scanning Probe Instruments Inc. (ICSPI) has had a MEMS-based AFM scanner on the market for a few years.[11] It is worth noting that ICSPI is making these MEMS AFMs in a CMOS foundry. We should also point out that the DOE has recently awarded two research contracts to develop MEMS STM scanners for the express purpose of engineering them to create arrays of MEMS STM scanners. We finally point out that these research programs while extremely worthwhile, would need to be followed up with significantly larger funded programs to develop manufacturing tools.

**References**

1    Frank J. Rueß, Wilson Pok, Thilo C. G. Reusch, Matthew J. Butcher, Kuan Eng J. Goh, Lars Oberbeck, Giordano Scappucci, Alex R. Hamilton, and Michelle Y. Simmons, *Realization of Atomically Controlled Dopant Devices in Silicon*, Small **3**, 563–567 (2007).

2   M. C. Hersam, N. P. Guisinger, and J. W. Lyding, *Silicon-based molecular nanotechnology*, Nanotechnology **11**, 70–76 (2000).

3   J. N. Randall, J. B. Ballard, J. W. Lyding, S. Schmucker, J. R. Von Ehr, R. Saini, H. Xu, Y. Ding, *Atomic precision patterning on Si: An opportunity for a digitized process*, Microelectron. Eng. **87**, 955–958 (2010).

4   V. R. Manfrinato, J. Wen, L. Zhang, Y. Yang, R. G. Hobbs, B. Baker, and K. K. Berggren, *Determining the resolution limits of electron-beam lithography: Direct measurement of the point-spread function*, Nano Letters **14**(8), 4406–4412 (2014). http://doi.org/10.1021/nl5013773

5   J. N. Randall, J. H. G. Owen, J. Lake, R. Saini, E. Fuchs, M. Mahdavi, S. O. Reza Moheimani, and B. C. Schaefer, *Highly parallel scanning tunneling microscope based hydrogen depassivation lithography*, Journal of Vacuum Science & Technology B **36**, 6–10 (2018). http://doi.org/10.1116/1.5047939

6   J. B. Ballard, T. W. Sisson, J. H. G. Owen, W. R. Owen, E. Fuchs, J. Alexander, J. N. Randall, and J. R. Von Ehr, *Multimode hydrogen depassivation lithography: A method for optimizing atomically precise write times*, J. Vac. Sci. Technol. B Microelectron. Nano. Struct. **31**, 06FC01-1 - 06FC01-6 (2013).

7   F. E. Kalff, M. P. Rebergen, E. Fahrenfort, J. Girovsky, R. Toskovic, J. L. Lado, J. Fernández-Rossier, and A. F. Otte, *A kilobyte rewritable atomic memory*, Nat. Nanotechnol. **11**, 926–929 (2016).

8   M. R. Slot, T. S. Gardenier, P. H. Jacobse, G. C. P. van Miert, S. N. Kempkes, S. J. M. Zevenhuizen, C. Morais Smith, D. Vanmaekelbergh, and I. Swart, *Experimental realization and characterization of an electronic Lieb lattice,* Nature Physics **13**, 672 (2017).

9   D. M. Tennant, *Nanotechnology.* (G. Timp, Ed.). New York: Springer-Verlag (1999).

10  D. M. Tennant, *Progress and issues in e-beam and other top down nanolithography*, Journal of Vacuum Science & Technology A: Vacuum, Surfaces, and Films **31**(5), 050813 (2013). http://doi.org/10.1116/1.4813761

11  https://www.icspicorp.com/

# 4. Theoretical limits, legacies, and prospects

## 4.1. *Theoretical limits and experimental prospects needed to study new physics*

Ehsan Khatami,[1] Richard Scalettar,[2] and Erik Henriksen[3]

[1]*San Jose State University,* [2]*University of California, Davis,* [3]*Washington University*

The Hubbard Hamiltonian is widely believed to model some of the most important and fundamental phenomena of correlated electron materials. At half-filling, the density, which is 'most metallic' from the viewpoint of band theory, the interaction $U$ instead causes Mott insulator behavior. At the same time, long range antiferromagnetic correlations emerge. In this way, the half-filled Hubbard Hamiltonian captures the antiferromagnetic insulating behavior of transition metal monoxides like MnO, a mineral of considerable geophysical interest.

When doped, the behavior is even more complex. Besides other types of magnetic ordering (ferromagnetism and incommensurate/striped phases), both d-wave superconductivity and charge ordering central to cuprate superconductors appear to be captured. Indeed, the Hubbard Hamiltonian in other geometries can exhibit exotic behavior which has been the focus of much of the forefront of condensed matter physics over the last decade — spin liquid and flat band physics on a Kagome lattice (herbertsmithite), the effect of a Dirac dispersion relation on a honeycomb lattice (graphene), and AF-singlet transitions in bilayer (two-band) geometries (heavy fermion systems and lanthanide volume collapse transitions), to name just a few of the many themes.

Despite the remarkable prospect of understanding this wide range of correlation physics qualitatively, our current ability to obtain accurate theoretical and computational results for the Hubbard Hamiltonian remains limited to quite restricted situations: one- and quasi-one-dimensional systems,[1] half filling,[2,3] the limit of infinite dimension,[4,5] or relatively high temperatures.[6,7] The key reason for this failure is that much of the interest is in the intermediate coupling regime ($U \sim$ non-interacting bandwidth) where analytic approaches fail. While both analytic and numerical methods can be made to work by introducing approximations, they are almost sure to affect the subtle interplay between the different phases and hence give inaccurate results.

Numerical approaches are likewise limited. The exponentially large Hilbert space constrains exact diagonalization to systems of only a dozen or so sites. Quantum Monte Carlo, which is extremely powerful for quantum spin and boson Hamiltonians, fails due to the "sign problem" — the quantity that plays the role of the probability of particular fermionic arrangements can go negative, precluding its use in generating configurations on which measurements can be performed.

A more detailed discussion of the challenges to a computational solution of the Hubbard Hamiltonian, including the strengths and weaknesses of specific approaches, is covered in Section II — Technical Addendum.

In view of these theoretical limitations, it is natural to ask what experimental prospects exist for elucidating the physics of models of correlated electrons. A similar question has been asked over the last decade in the field of ultracold atoms trapped in optical lattices, where it was suggested that the Hubbard Hamiltonian with high tunability of parameters might be realized without many of the complexities of real materials. Tremendous progress has been made. Unfortunately, cooling fermionic atoms has proved to be very challenging, and only at great effort have temperatures on the order of the exchange constant been reached. Thus, while short range AF order has been realized, the Néel transition remains elusive. The d-wave superconducting transition, if indeed it is present, remains well out of reach, a factor of 4 to 5 yet lower in temperature. Theory/experiment collaborations in this field[8-14] not only have led to better characterization of, and new discoveries with, cold atom systems, but also promoted the development and refinement of several numerical methods, such as QMC, dynamical mean field theory, finite-temperature Lanczos techniques, and numerical lined-cluster expansions, which can now tackle dynamical properties of the Hubbard model.[13,14]

**Experimental Prospects**

Arrays of hundreds or perhaps thousands of dopant atoms in silicon are technologically feasible. This presents opportunities to precisely engineer meta-materials for the purpose of better understanding Hubbard model physics, which is "built in" to the device architecture. Such metamaterials are limited in size to a few microns or so, but this readily crosses the threshold to where they may be treated as any other mesoscopic device, and therefore can be experimentally studied using a wide range of tools and techniques already existing for work with "real" materials. The literature on such measurements is vast and no attempt to sum it up here will be made; rather, we briefly outline several potential approaches for studying such dopant arrays.

The most straightforward approach, even when expecting Mott-like insulating behavior, is to attach metallic leads and measure the electronic transport at low frequencies. Such leads can be readily incorporated into and on the top of the Si wafer hosting the dopant array. One may imagine simple two- or four-terminal transport measurements that will readily determine the system conductivity irrespective of shape via van der Pauw measurements; the charge carrier density via Hall measurements in weak magnetic fields; or the gap to the next conducting band by activation measurements over a range of temperatures. Indeed, promising first steps in these directions have been taken.[15] In larger devices enabling multi-terminal geometries, "non-local" measurements in which the current and voltage leads are well-separated can reveal the presence of long-range, non-Ohmic transport of the type encountered in quantum Hall effects or when the spin and charge degrees of freedom diffuse independently.[16,17] These effects may arise as a consequence of novel band structure topology or reveal significantly enhanced Zeeman splitting.

Scanned probe techniques are expected to play a major role in exploring dopant arrays. While the dopants are likely buried too deep to be accessed by scanning tunneling microscopy, far-field probes such as scanning gate or scanning Kelvin probe microscopy may be employed. In the former, transport is measured through contacts to the device as a function of the potential applied to a moveable metal tip.[18] As the tip couples capacitively to the dopant array, transport will reflect changes in the bulk due to localized doping induced by the distant tip. In contrast, scanning Kelvin probe microscopy determines the electric potential difference between tip and sample, enabling extraction of the work function if there is no applied bias. The likely most useful application to dopant arrays is to perform Kelvin probe microscopy when a local metallic gate is buried below the dopant array. Such a gate can be used to capacitively vary the chemical potential of the dopant array, changing its carrier density. If an AC modulation is applied to the back gate, the Kelvin probe will detect the change in electric potential in response; this measurement modality enables a direct measurement of the thermodynamic density of states, $\partial n/\partial \mu$ (also known as the inverse compressibility).[19]

Optical methods as well can be powerful probes of dopant arrays. Due to their small size, photoconductivity approaches are likely to be fruitful, where the electronic transport of the array is monitored as above while the sample is illuminated at light frequencies of interest. In particular, Fourier-transform spectroscopy can reveal the energy-dependent response when the sample is illuminated by a broadband light source; further information can be gained by shining the incident light through a grating to pick out particular wavevectors.[20] The sample response is generally due to both photo-excitation and bolometric effects. At much lower energies in the microwave regime, these dopant array devices may be incorporated into a microwave transmission line. For instance, a coplanar waveguide may be fabricated on the Si wafer surface, configured so that the array provides a path to ground. Measurement of the microwave transmission or reflection through this circuit will directly reveal the device impedance and how it changes as a function of, e.g., temperature and applied electric or magnetic fields.[21]

The techniques briefly enumerated here are far from exhaustive. Indeed, one of the very desirable features of these dopant array metamaterials is their outward resemblance to "real" materials and devices and the concomitant ability to probe them using many standard, off-the-shelf experimental probes. With the ability to tailor the Hubbard interactions via dopant spacing and arrangement, significant opportunities exist to experimentally access physics of the Hubbard model.

**References**

1    S. R. White, *Spin Gaps in a Frustrated Heisenberg Model for $CaV_4O_9$*, Phys. Rev. Lett. **77**, 3633 (1996).

2    R. Blankenbecler, D. J. Scalapino, and R. L. Sugar, *Monte Carlo calculations of coupled boson-fermion systems I.*, Phys. Rev. D, **24**, 2278 (1981).

3    C. N. Varney, C.-R. Lee, Z. J. Bai, S. Chiesa, M. Jarrell, R. T. Scalettar, *Quantum Monte Carlo study of the two-dimensional fermion Hubbard model*, Phys. Rev. B **80**, 075116 (2009).

4    M. Jarrell, *Hubbard model in infinite dimensions: A quantum Monte Carlo study*, Phys. Rev. Lett. **69**, 168 (1992).

5    A. Georges and G. Kotliar, *Hubbard Model in Infinite Dimensions*, Phys. Rev. B **45**, 6479 (1992).

6    J. Oitmaa *et al.*, *Series expansion methods for strongly interacting lattice models* (Cambridge University Press, Cambridge, 2006).

7    E. Khatami and M. Rigol, *Thermodynamics of strongly interacting fermions in two-dimensional optical lattices*, Phys. Rev. A **84**, 053611 (2011).

8    R. A. Hart *et al.*, *Observation of antiferromagnetic correlations in the Hubbard model with ultracold atoms*, Nature **519**, 211–214 (2015).

9    L. W. Cheuk *et al.*, *Observation of spatial charge and spin correlations in the 2D Fermi-Hubbard model*, Science **353**, 1260–1264 (2016).

10   A. Mazurenko *et al.*, *A cold-atom Fermi–Hubbard antiferromagnet*, Nature **545**, 462–466 (2017).

11   P. T. Brown *et al.*, *Spin-imbalance in a 2D Fermi-Hubbard system*, Science **357**, 1385–1388 (2017).

12   D. Mitra *et al.*, *Quantum gas microscopy of an attractive Fermi-Hubbard system*, Nature Physics **14**, 173–177 (2017).

13   P. T. Brown *et al.*, *Bad metallic transport in a cold atom Fermi-Hubbard system*, arXiv:1802.09456 (2018).

14   M. A. Nichols *et al.*, *Spin Transport in a Mott Insulator of Ultracold Fermions*, arXiv:1802.10018 (2018).

15  E. Prati *et al., Band transport across a chain of dopant sites in silicon over micron distances and high temperatures*, Scientific Reports, **6**, 19704 (2016).

16  A. Roth *et al., Nonlocal Transport in the Quantum Spin Hall State*, Science 325, 294 (2009).

17  D. Abanin *et al., Giant nonlocality near the Dirac point in graphene*, Science **332**, 328 (2011).

18  H. Sellier *et al., On the imaging of electron transport in semiconductor quantum structures by scanning-gate microscopy: successes and limitations*, Semiconductor Science and Technology **26**, 064008 (2011).

19  J. Martin *et al., Observation of electron–hole puddles in graphene using a scanning single-electron transistor*, Nature Physics **4**, 144 (2008).

20  S. Holland *et al., Quantized dispersion of two-dimensional magnetoplasmons detected by photoconductivity spectroscopy*, Phys. Rev. Lett. **93**, 186804 (2004).

21  P. Jiang *et al., Quantum oscillations observed in graphene at microwave frequencies*, Appl. Phys. Lett. **97**, 062113 (2010).

## 4.2. Types of physics involved and kinds of problems that might be solved

### 4.2.1. Fermi–Hubbard model: its importance and its role

Subir Sachdev

*Harvard University*

The Hubbard model was originally introduced as a model of the metal-insulator transition driven by correlation energies rather than Anderson localization from disorder. Hubbard's insight was that one could focus just on the on-site repulsion $U$, as the longer-range part of the Coulomb interaction is subject to screening, at least in the metallic phase. And a model with just an on-site repulsion is sufficient to create an insulating state with a large energy gap at sufficiently large $U$. And a strong-coupling expansion from within the localized metallic phase also becomes possible in a Hubbard-like model.

The initial applications of the Hubbard model in the 1970's and 80' were to doped semiconductors, especially at doping densities close to the metal-insulator transition. Many models were developed for the role of spin fluctuations near the metal-insulator transition, and these led to some understanding of the relative independence of the spin and charge sectors *i.e.* the spin susceptibility shows little change, while the charge sector changes from metallic to insulating.

With the discovery of the cuprate superconductors, interest in the Hubbard model exploded, after Anderson's proposal that it could also explain the essential physics of high-temperature superconductivity. This hope has generally been borne out in subsequent years, although many aspects have turned out to be far more complicated than anticipated.

With advent of microscopic control of electronic degrees of freedom, the Fermi–Hubbard model has continued to be of significant importance. Now we are interested in effects of quantum entanglement between different sites of the Hubbard model, and its consequences on dynamics. New theoretical tools are being developed to understand such regimes.

4.2.2.  *What can the Fermi–Hubbard model teach about metal-insulator transitions, magnetism, exotic superconductivity, and materials like transition metal monoxides, cuprates, heavy fermions?*

Vladimir Dobrosavljevic[1] and Gabe Aeppli[2]

[1]*Florida State University,* [2]*Paul Scherrer Institute, Switzerland*

**Metal-Insulator Transitions**

Metals and insulators are very different states of matter. Both, however, are characterized by stable and robust electronic states, among which the electrons rearrange themselves, as one varies the temperature or applies external fields. This simplicity affords excellent—textbook grade—understanding of these materials, but it also limits our ability to "tweak" them, *i.e.* to significantly alter their behaviors through modest perturbations. Much more interesting are materials that find themselves somewhere in between. In this Metal-Insulator Transition (MIT) region,[1,2] spectacular response is observed with modest tuning of control parameters, often displaying "Strange" or "Bad Metal" behavior.[3-5] In these systems, which include the familiar (yet still poorly understood) doped semiconductors, but also various transition metal oxides and many other families of correlated matter, the appealing physical picture of nearly-free electrons fail in dramatic ways. Here one observes dramatic evolution and modification of the electronic spectra with modest variation of temperature, pressure, doping or— what is of most technological relevance—gating, indicating pronounced many-body effects. The task to understand and to describe how the electronic states respond and adjust, as one tunes the system from insulator to metal, this is a basic science challenge of our era.

**A. Anderson vs. Mott: The long-standing mystery of Si:P**

Phosphorus doped silicon (Si:P) is one of the first and perhaps the best known[6,7] material manifesting a well-characterized MIT,[8,9] without displaying any change of symmetry across the transition. Indeed, no evidence for spin, charge, orbital, structural, or even spin-glass order has been detected in its vicinity, leaving only the strong electron correlations and/or disorder as potential key players. The MIT in this bellwether material, which is representative of a large class of doped semiconductors,[10] stands prominent as one of the key unresolved problems[11] in solid state physics. Renewed interest in Si:P (and related materials) returned in full force in the last several years,[12-16] in part because of its possible relevance to quantum computing. This led to considerable advance of various experimental capabilities to fabricate and precisely characterize such metamaterials on a nanoscale,[12-16] opening a path to new understanding of these systems on the fundamental level, as well as for many possible applications.

The following well-established features of Si:P stand out: (1) Each phosphorus (P) donor binds exactly one electron at low density,[17] forming a spin-½ local magnetic moment; the low density phase can, therefore, be regarded as a strongly disordered Mott insulator.[18] (2) Due to their random placement, each donor spin forms a spin dimer[19] with one preferred neighbor; the resulting random singlet phase,[19,20] features characteristic power-law singularities for the spin susceptibility and the Sommerfeld coefficient ($\chi, \gamma \sim T^{-\alpha}$, with an effective exponent $\alpha \approx 0.7$). (3) Such thermodynamic singularities persist[21] very deep into the metallic phase, thus displaying disorder-induced non-Fermi-liquid behavior.[22-24] Remarkably, a thermodynamic response alone cannot even be utilized to locate the transition. (4) In contrast, low-temperature transport displays a very sharply manifested MIT,[8,9] with conductivity $\sigma \sim (n - n_C)^\mu$, where $\mu \approx 0.5$ for essentially all known doped semiconductors.[7,25] It should be stressed that this exponent is much smaller (so the transition is much more "abrupt"—almost a jump!) than what is expected for (non-interacting) Anderson localization ($\mu \approx 1.6$), or even classical percolation ($\mu \approx 2$). This body of evidence strongly suggests that one should not think of MIT in Si:P as an Anderson-like transition, but instead as a disordered Mott transition. But what kind of Mott transition is it?

**B. Hubbard–Mott vs. "charge-transfer": Two faces of the Mott transition**

The conceptually simplest picture of Mott transition is that of a narrow half-filled band,[26] where sufficiently strong on-site Coulomb repulsion opens a gap at the Fermi energy. This scenario proves sufficient in systems such as κ-organics,[27,28] where a single-band Hubbard model suffices.[29] More complicated situations are also well known, including many charge-transfer (transition-metal) oxides, as first realized by Zaanen, Sawatzky, and Allen,[30] and also various actinide oxides,[31] *etc.* In all these systems, "orbital-selective" Mott localization may involve only some of the electrons, and the delocalization on the metallic side typically involves significant charge transfer,[30] to other bands/orbitals.[32]

Remarkably, this "charge-transfer" mechanism for the Mott transition, was anticipated in Mott's pioneering work on doped semiconductors, dating back to 1949.[33] At the lowest densities, isolated donors form singly occupied bound states with 1s character. At metallic densities, however, the screening effects should significantly diminish the attractive potential of the donor ions, completely destroying the bound states and the associated impurity band! Indeed, Mott provided simple heuristic arguments,[33] suggesting that as a consequence, the transition in the clean limit should assume a first-order character. With few exceptions,[34] this physical mechanism has largely been ignored over the years, and most theoretical work focused on solving Hubbard-type models,[18-20] implicitly brushing aside the potentially significant role of the long-range Coulomb interactions, and the associated role of "full charge self-consistency."[35] Given the long history of unsuccessful theoretical efforts [11,36] to understand and describe the MIT in Si:P, it may be time for serious re-evaluation of the key mechanisms involved. Unfortunately, strong positional disorder, which is generically found in bulk doped semiconductors, creates an added layer of complexity; it now becomes difficult to disentangle the role of disorder from the strong correlation effects, making the problem almost hopelessly unsolvable.

## C. The new beginning: Si:P "metamaterials"

A very important new opportunity to shed light on this age-old puzzle is provided by very recent experimental advances[13,15] that allow engineering, with atomic precision, any desired pattern of donor ions on the surface of a semiconductor. These new capabilities, which were initially stimulated by efforts[12,14,16] to utilize isolated donor ions as qubits in a quantum computer, should allow precise and careful addressing and answering various basic questions of fundamental importance, especially if one is able to couple these experimental efforts with careful and quantitative theory modeling for each specific situation.

As a first step, one should create periodic arrays of donor ions and vary the lattice spacing, to tune the system through a Mott transition. In this way, one may be able to first unravel the nature of the corresponding Mott transition in the absence of randomness. Theoretical studies should be carried out in direct collaboration with experimental groups, such as that of Gabriel Aeppli (Paul Scherrer Institute ETH Zurich and EPF Lausanne) as well as his colleagues in London (Schofield and Curson) and Surrey (Murdin), who will be doing transport measurements as well as various photon-based probes, with wavelengths from microwaves to soft x-rays, which will allow angle-resolved photoemission from (buried) delta layers. The theory component should be performed using the state-of-the art LDA+RISB[31,37] and/or LDA+DMFT methods,[38] which have been recently developed and validated for other systems. Within this setup, the following important questions can be addressed:

1.  What will be the consequences of the charge-transfer mechanism at the MIT? The self-consistent modification of screening is tackled at the static (Hartree) level within LDA+RISB or LDA+DMFT, since we need to determine if this is sufficient to drive the transition first order, as Mott anticipated.[33]

2.  Dynamic screening may also play a significant role, as it can dramatically reduce the magnitude of the on-site Coulomb repulsion $U$ on the metallic side. This effect, which has recently been much discussed[39] for example in the context of iron-pnictides, can be incorporated by using $GW$[40] and EDMFT methods.[41,42]

3. Another important ingredient that is left out in the simplest "single-site" DMFT implementations is the role of inter-site spin correlations. This effect is especially important in the presence of dimerization, found in $VO_2$, and as established in recent work;[43,44,45] thus, providing another mechanism to strengthen the first-order character of the Mott transition. The role of dimerization is expected to be even more significant in doped semiconductors, where positional randomness induces the formation of spin dimers on the insulating side. To study this effect in the absence of disorder, one should first introduce dimerization in a periodic lattice of donor ions, both experimentally and theoretically.

4. In presence of disorder, the anticipated first-order nature of the MIT would immediately trigger nano-scale phase separation[46] between the metallic and the insulating domains, providing a clear immediate mechanism for the well-documented "two-fluid" thermodynamic behavior in doped semiconductors.[21] The additional effects of disorder could also be directly tested experimentally (see Fig. 3), by systematically introducing randomness in the spatial arrangements of the donor atoms "written" by STM tips. The resulting charge transfer should also make it possible to directly detect by available scanning microwave microscopy imaging on the nano-scale, for any donor geometry one decides to examine.[15] The corresponding theoretical modeling should be carried out using, *e.g.*, the methods outlined above.

The described progress on both the experimental and the theoretical front is opening a clear path towards a long-awaited solution of the MIT puzzle in doped semiconductors. From a broader perspective, these ideas and methods should be of consequence in many other systems, for example the underdoped regime of various Mott oxides, including not only the superconducting cuprates[47] but also the equally interesting iridium oxide materials such as $(Sr_{1-x}La_x)_2IrO_4$, which display similar physics puzzles.[48,49]

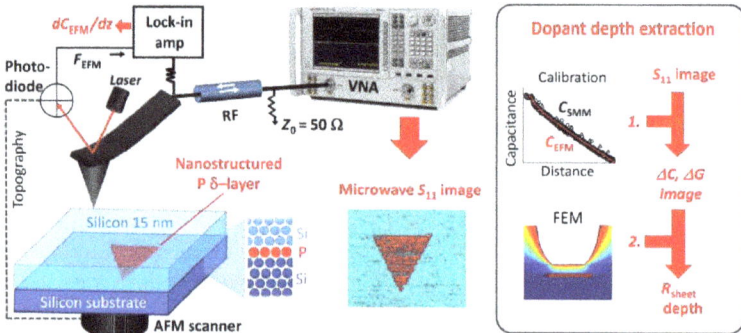

Fig. 3. Experimental setup for imaging the "buried" arrays of donor ions, which can be STM-implanted under the surface of a semiconductor.[15] This capability allows not only very precise control in designing patterns with desired concentration and/or geometry, but also imaging of the corresponding electronic states on the nano-scale; thus revealing, the charge rearrangements across the metal-insulator transition, with or without randomness.

## References

1　N. F. Mott, *Metal-Insulator Transition* (Taylor & Francis, London, 1990).

2　V. Dobrosavljević, N. Trivedi, and J. M. Valles Jr, *Conductor Insulator Quantum Phase Transitions* (Oxford University Press, UK, 2012).

3　H. Takagi, B. Batlogg, H. L. Kao, J. Kwo, R. J. Cava, J. J. Krajewski, and W. F. Peck, *Systematic evolution of temperature-dependent resistivity in $La_{2-x} Sr_x CuO_4$*, Phys. Rev. Lett. **69**, 2975 (1992).

4　V. J. Emery and S. A. Kivelson, *Superconductivity in bad metals*, Phys. Rev. Lett. **74**, 3253 (1995).

5　N. E. Hussey, K. Takenaka, and H. Takagi, *Universality of the Mott-Ioffe-Regel limit in metals*, Philos. Mag. **84**, 2847 (2004).

6　M. A. Paalanen and R. N. Bhatt, *Transport and thermodynamic properties across the metal-insulator transition*, Physica B **169**, 231 (1991).

7　M. P. Sarachik, in P. Edwards and C. N. R. Rao, editors, *Metal-Insulator Transitions Revisited* (Taylor and Francis, London, 1995).

8　T. Rosenbaum, K. Andres, G. Thomas, and R. Bhatt, *Sharp metal-insulator transition in a random solid*, Phys. Rev. Lett. **45**, 1723 (1980).

9　M. A. Paalanen, T. F. Rosenbaum, G. A. Thomas, and R. N. Bhatt, *Stress tuning of the metal-insulator transition at millikelvin temperatures*, Phys. Rev. Lett. **48**, 1284 (1982).

10　B. Shklovskii and A. Efros, *Electronic properties of doped semiconductors*, Springer series in Solid-State Sciences (Springer-Verlag, Berlin, 1984).

11　P. A. Lee and T. V. Ramakrishnan, *Disordered electronic systems*, Rev. Mod. Phys. **57**, 287 (1985).

12　P. T. Greenland, S. A. Lynch, A. F. G. van der Meer, B. N. Murdin, C. R. Pidgeon, B. Redlich, N. Q. Vinh, and G. Aeppli, *Coherent control of Rydberg states in silicon*, Nature **465**, 1057 (2010).

13　S. R. Schofield, P. Studer, C. F. Hirjibehedin, N. J. Curson, G. Aeppli, and D. R. Bowler, *Quantum engineering at the silicon surface using dangling bonds*, Nature Communications **4**, 1649 (2013).

14　K. L. Litvinenko, E. T. Bowyer, P. T. Greenland, N. Stavrias, J. Li, R. Gwilliam, B. J. Villis, G. Matmon, M. L. Y. Pang, B. Redlich, A. F. G. van der Meer, C. R. Pidgeon, G. Aeppli, and B. N. Murdin, *Coherent creation and destruction of orbital wavepackets in Si:P with electrical and optical read-out,* Nature Communications **6**, 6549 (2015).

15　G. Gramse, A. Kölker, T. Lim, T. J. Z. Stock, H. Solanki, S. R. Schofield, E. Brinciotti, G. Aeppli, F. Kienberger, and N. J. Curson, *Nondestructive imaging of atomically thin nanostructures buried in silicon*, Science Advances **3,** (2017).

16　S. Chick, N. Stavrias, K. Saeedi, B. Redlich, P. T. Greenland, G. Matmon, M. Naftaly, C. R. Pidgeon, G. Aeppli, and B. N. Murdin, *Coherent superpositions of three states for phosphorous donors in silicon prepared using THz radiation*, Nature Communications **8**, 16038 (2017).

17 G. A. Thomas, M. Capizzi, F. DeRosa, R. N. Bhatt, and T. M. Rice, *Optical study of interacting donors in semiconductors*, Phys. Rev. B **23**, 5472 (1981).

18 M. Milovanović, S. Sachdev, and R. N. Bhatt, *Effective-field theory of local-moment formation in disordered metals*, Phys. Rev. Lett. **63**, 82 (1989).

19 R. N. Bhatt and P. A. Lee, *Scaling studies of highly disordered Spin-1/2 anti-ferromagnetic systems*, Phys. Rev. Lett. **48**, 344 (1982).

20 S. Zhou, J. A. Hoyos, V. Dobrosavljevic, and E. Miranda, *Valence-bond theory of highly disordered quantum antiferromagnets,* Europhysics Letters **87**, 27003 (2009).

21 M. A. Paalanen, J. E. Graebner, R. N. Bhatt, and S. Sachdev, *Thermodynamic behavior near a metal-insulator transition*, Phys. Rev. Lett. **61**, 597 (1988).

22 G. R. Stewart, *Non-Fermi-liquid behavior in d- and f-electron metals*, Rev. Mod. Phys. **73**, 797 (2001).

23 E. Miranda and V. Dobrosavljevic, *Disorder-driven non-Fermi liquid behavior of correlated electrons*, Reports on Progress in Physics **68**, 2337 (2005).

24 E. C. Andrade, A. Jagannathan, E. Miranda, M. Vojta, and V. Dobrosavljević, *Non-Fermi- liquid behavior in metallic quasicrystals with local magnetic moments*, Phys. Rev. Lett. **115**, 036403 (2015).

25 M. P. Sarachik, D. Simonian, S. V. Kravchenko, S. Bogdanovich, V. Dobrosavljevic, and G. Kotliar, *Metal-insulator transition in Si:X (X=P,B): Anomalous response to a magnetic field*, Phys. Rev. B **58**, 6692 (1998).

26 J. Hubbard, *Electron correlations in narrow energy bands*, Proc. R. Soc. (London) A **276**, 238 (1963).

27 P. Limelette, P. Wzietek, S. Florens, A. Georges, T. A. Costi, C. Pasquier, D. Jerome, C. Meziere, and P. Batail, *Mott transition and transport crossovers in the organic compound*, Phys. Rev. Lett. **91**, 016401 (2003).

28 T. Furukawa, K. Miyagawa, H. Taniguchi, R. Kato, and K. Kanoda, *Quantum criticality of Mott transition in organic materials*, Nature Physics **11**, 221 (2015).

29 V. Dobrosavljević and D. Tanasković, *Wigner-Mott quantum criticality: from 2D-MIT to $^3$He and Mott organics*, in S. Kravchenko, editor, Strong Correlation Phenomena around 2D Conductor-Insulator Transitions (Pan Stanford Publishing, 2016).

30 J. Zaanen, G. A. Sawatzky, and J. W. Allen, *Band gaps and electronic structure of transition-metal compounds*, Phys. Rev. Lett. **55**, 418 (1985).

31 N. Lanatà, Y. Yao, X. Deng, V. Dobrosavljević, and G. Kotliar, *Slave boson theory of orbital differentiation with crystal field effects: Application to $UO_2$*, Phys. Rev. Lett. **118**, 126401 (2017).

32    T.-H. Lee, Y.-X. Yao, V. Stevanović, V. Dobrosavljević, and N. Lanatà, *Critical role of electronic correlations in determining crystal structure of transition metal compounds*, arXiv:1710.08586, submitted for publication to Nature Physics (2017).

33    N. F. Mott, Proc. Roy. Soc. (London) A **197**, 269 (1949).

34    R. Chitra and G. Kotliar, *Effect of long-range Coulomb interactions on the Mott transition*, Phys. Rev. Lett. **84**, 3678 (2000).

35    K. Haule, C.-H. Yee, and K. Kim, *Dynamical mean-field theory within the full-potential methods: Electronic structure of CeIrIn₅, CeCoIn₅, and CeRhI₅*, Phys. Rev. B **81**, 195107 (2010).

36    D. Belitz and T. R. Kirkpatrick, *The Anderson-Mott transition*, Rev. Mod. Phys. **66**, 261 (1994).

37    N. Lanatà, T.-H. Lee, Y. Yao, and V. Dobrosavljević, *Emergent Bloch excitations in Mottmatter*, Phys. Rev. B **96**, 195126 (2017).

38    G. Kotliar, S. Y. Savrasov, K. Haule, V. S. Oudovenko, O. Parcollet, and C. A. Marianetti, *Electronic structure calculations with dynamical mean-field theory*, Rev. Mod. Phys. **78**, 865 (2006).

39    A. van Roekeghem, L. Vaugier, H. Jiang, and S. Biermann, *Hubbard interactions in iron-based pnictides and chalcogenides: Slater parametrization, screening channels, and frequency dependence*, Phys. Rev. B **94**, 125147 (2016).

40    A. Kutepov, K. Haule, S. Y. Savrasov, and G. Kotliar, *Self-consistent GW determination of the interaction strength: Application to the iron arsenide superconductors*, Phys. Rev. B **82**, 045105 (2010).

41    Y. Pramudya, H. Terletska, S. Pankov, E. Manousakis, and V. Dobrosavljević, *Nearly frozen Coulomb liquids*, Phys. Rev. B **84**, 125120 (2011).

42    F. Nilsson, L. Boehnke, P. Werner, and F. Aryasetiawan, *Multitier self-consistent GW + EDMFT*, Phys. Rev. Materials **1**, 043803 (2017).

43    G. Moeller, V. Dobrosavljević, and A. E. Ruckenstein, *RKKY interactions and the Mott transition*, Phys. Rev. B **59**, 6846 (1999).

44    O. Nájera, M. Civelli, V. Dobrosavljević, and M. J. Rozenberg, *Resolving the VO₂ controversy: Mott mechanism dominates the insulator-to-metal transition*, Phys. Rev. B **95**, 035113 (2017).

45    O. Nájera, M. Civelli, V. Dobrosavljević, and M. J. Rozenberg, *Multiple crossovers and coherent states in a Mott-Peierls insulator*, arXiv:1707.09310, submitted for publication to Phys. Rev. B (2017).

46    E. Dagotto, Nanoscale Phase Separation and Colossal Magnetoresistance (Springer-Verlag, Berlin, 2002).

47    M. A. Sulangi, M. P. Allan, and J. Zaanen, *Revisiting quasiparticle scattering interference in high-temperature superconductors: The problem of narrow peaks*, Phys. Rev. B **96**, 134507 (2017).

46

48   I. Battisti, K. M. Bastiaans, V. Fedoseev, A. de la Torre, N. Iliopoulos, A. Tamai, E. C. Hunter, R. S. Perry, J. Zaanen, F. Baumberger, and M. P. Allan, *Universality of pseudogap and emergent order in lightly doped Mott insulators*, Nature Physics **13**, 21 (2016).

49   I. Battisti, V. Fedoseev, K. M. Bastiaans, A. de la Torre, R. S. Perry, F. Baumberger, and M. P. Allan, *Poor electronic screening in lightly doped Mott insulators observed with scanning tunneling microscopy*, Phys. Rev. B **95**, 235141 (2017).

### 4.3. Legacy, limitations, and prospects in view of strongly correlated Hamiltonians, including the Fermi-Hubbard model

#### 4.3.1. Via precisely atomically-doped semiconductors

Subir Sachdev

*Harvard University*

Modern ideas on correlated quantum states should give a new perspective on older experiments on Si:P and would be valuable for understanding new structures that can be fabricated with atomic precision. Among early observations on Si:P near the metal-insulator transition was one that showed the spin susceptibility had a divergent form as the temperature $T \to 0$; and in addition, this divergence continued smoothly across the metal-insulator transition. This is very suggestive of spin-charge separation, in the presence of inhomogeneity. In more precise terms, we can imagine that both the insulating and metallic states have topological order and require emergent gauge fields for their description. Such exotic phenomena can have observable consequences for the thermal conductivity, and lead to dramatic effects on the nature of the metal-insulator transition. Although such theories of spin-charge separated states are hard to treat analytically, considerable progress has recently been possible in the Sachdev–Ye–Kitaev (SYK) models. We will extend the SYK-type analysis to more realistic situations found in Si:P. The ability to make atomically precise structures will enable tests of the SYK analysis and extend its predictions in the vicinity of the metal-insulator transition.

### 4.3.2. *Via optical lattices and cold atoms*

Kaden Hazzard and Bhuvanesh Sundar

*Rice University*

Ultracold matter already has greatly impacted our understanding of strongly correlated quantum matter, as mentioned in Sec. I (2.2). As a still-young and expanding field, it is too soon to write its "legacy," and as a broad field it is not possible to even survey the full landscape of results. Instead, we will focus attention on a line of research that uses repulsive fermionic atoms in optical lattices to realize the repulsive Fermi–Hubbard model.[a] Even though this represents a tiny fraction of the exciting frontier of quantum simulation in ultracold matter research, it provides an illustrative example of the character of much research in ultracold systems into strongly correlated matter. We provide a sketch here, with more detail given in Sec. II — Technical Addendum.

Even in the slice of the field concentrating on fermions in optical lattices, there have been exciting advances. Landmark experiments observed the metallic phase, band insulator,[1] and the Mott insulator,[2,3] as well as the doping-tuned Mott-metal crossover (at temperatures above magnetic ordering). These were first observed in the 3D Fermi–Hubbard model, but have since been seen in 2D and 1D. [More discussion and citations are provided in Sec. II — Technical Addendum.] In the last few years, experiments have achieved lower temperatures and advanced detection techniques, and have observed short-ranged anti-ferromagnetic correlations in 3D and long-ranged correlations spanning the system size ($\sim 20$ sites[4]) in 1D and 2D for the half-filling Fermi–Hubbard model. The current experimental temperatures, however, remain slightly above the Néel temperature $T_N$ at which the phase transition from the normal phase to the antiferromagnetic phase occurs, by about 40% in 3D, while in 1D and 2D the Néel temperature is $T_N = 0$. Nevertheless, even these experimental regimes are at the cusp of what is tractable theoretically. Broadly speaking with great effort, theory has been able to reproduce the equilibrium experimental results, thanks to significant advances in computational methods — exact diagonalization,[5] several quantum Monte Carlo (QMC) variants,[6-8] density matrix renormalization group (DMRG),[9,10] dynamical mean field theory (DMFT),[11] and numerical linked

---

[a] In solid-state physics, this is typically just referred to as the "Hubbard model," but in ultracold matter we routinely deal with bosons and fermions, so the phrase "Fermi–Hubbard model" has emerged as the standard name. We also routinely deal with both repulsive and attractive interactions; hence the name is extended further as the "repulsive Fermi–Hubbard model."

cluster expansion (NLCE) techniques[12] have all been influential in ultracold matter — which have been driven in no small part by these experiments. However, often the systematic errors such as finite size effects of these methods are difficult to assess in this regime, and the experiments offer some of the most stringent tests of their convergence.

Prospects for the future are even more exciting. With modest decreases in temperature, the Néel phase and the phase transition properties will be accessible, and — in the system doped from half filling — even equilibrium physics lies beyond the reach of well-controlled, quantitative theory. At the lower temperatures in the doped system, the putative existence of a $d$-wave superconducting phase could be studied, as could the phenomenology of any pseudogap and bad metal regimes of the model.[14,13]

However, challenges remain to reach the temperature required for these future prospects. Lower temperatures require removing entropy from the system, which is not possible in a perfectly closed system. Evaporative cooling reduces entropy and is the method used to bring atoms to ultracold temperatures, but in present conditions the cooling power provided by evaporative cooling is often balanced by intrinsic heating rates due to scattering from laser light and inelastic atomic collisions. To reach colder temperatures, the community must figure out ways to reduce the heating or find methods with sufficient cooling power to remove entropy before more is generated. Devising such methods is an outstanding goal and connects to deep physics, since understanding equilibration of these systems requires understanding transport and dynamics in strongly correlated systems.

Whatever methods are employed, current limits do not appear to be fundamental, and the history of cold matter suggests that major advances are still possible. As the challenges are understood and overcome, ultracold matter will impact science in even broader and deeper ways — the advances we can foresee will provide greater insight into strongly correlated matter, with far-reaching consequences for quantum materials, but the unforeseen advances are sure to be the most exciting.

## References

1 M. Köhl, H. Moritz, T. Stöferle, K. Günter, and T. Esslinger, *Fermionic atoms in a three-dimensional optical lattice: Observing Fermi surfaces, dynamics, and interactions*, Phys. Rev. Lett. **94**, 080403 (2005).

2 U. Schneider, L. Hackermüller, S. Will, T. Best, I. Bloch, T. A. Costi, R. W. Helmes, D. Rasch, and A. Rosch, *Metallic and insulating phases of repulsively interacting fermions in a 3D optical lattice*, Science **322**, 1520 (2008).

3   R. Jördens, N. Strohmaier, K. Günter, H. Moritz, and T. Esslinger, *A Mott insulator of fermionic atoms in an optical lattice*, Nature **455**, 204 (2008).

4   A. Mazurenko, C. S. Chiu, G. Ji, M. F. Parsons, M. Kanász-Nagy, R. Schmidt, F. Grusdt, E. Demler, D. Greif, and M. Greiner, *A cold-atom Fermi–Hubbard antiferromagnet*, Nature **545**, 462 (2017).

5   A. W. Sandvik, *Computational studies of quantum spin systems*, AIP Conference Proceedings **1297**, 135 (2010).

6   L. Pollet, *Recent developments in quantum Monte Carlo simulations with applications for cold gases*, Reports on Progress in Physics **75**, 094501 (2012).

7   L. K. Wagner and D. M. Ceperley, *Discovering correlated fermions using quantum Monte Carlo*, Reports on Progress in Physics **79**, 094501 (2016).

8   R. Blankenbecler, D. J. Scalapino, and R. L. Sugar, *Monte Carlo calculations of coupled boson-fermion systems. I*, Phys. Rev. D **24**, 2278 (1981).

9   S. R. White, *Density matrix formulation for quantum renormalization groups*, Phys. Rev. Lett. **69**, 2863 (1992).

10  U. Schollwöck, *The density-matrix renormalization group in the age of matrix product states*, Annals of Physics **326**, 96 (2011).

11  A. Georges, G. Kotliar, W. Krauth, and M. J. Rozenberg, *Dynamical mean-field theory of strongly correlated fermion systems and the limit of infinite dimensions*, Rev. Mod. Phys. **68**, 13 (1996).

12  B. Tang, E. Khatami, and M. Rigol, *A short introduction to numerical linked-cluster expansions*, Computer Physics Communications **184**, 557 (2013).

13  O. Gunnarsson, M. Calandra, and J. E. Han, *Colloquium: Saturation of electrical resistivity*, Rev. Mod. Phys. **75**, 1085 (2003).

### 4.3.3. *Via trapped ions*

Norbert Linke

*University of Maryland*

Control over individually trapped ions has developed rapidly in the past two decades. Thanks to their unmatched coherence properties,[1] and long-range coupling *via* the Coulomb interaction,[2] trapped ions have become a versatile platform employed for a large variety of experimental goals. The steady stream of ground-breaking results reaches from quantum simulation in the realm of quantum many-body phenomena,[3,4] over quantum chemistry studies including the control of all degrees of freedom in molecules,[5] to some of the most precise atomic clocks ever built.[6] Trapped ions are also one of the leading contenders for realizing a universal circuit model quantum computer.[7,8]

A good candidate system to create a quantum computer or quantum simulator has to fulfill a set of requirements such as the ability to perform state initialization, a set of operations, and read-out.[9,10] Additionally, in order to be useful, the quantum system needs to have low or well controlled coupling to the environment in order to exhibit coherence times that are long compared to the evolution under investigation. Furthermore, we need to be able to control sufficient degrees of freedom to usefully model a quantum problem, *i.e.*, in a quantum computer this corresponds to being able to scale to a sufficiently large number of qubits.

A single atom trapped in isolation from any other atoms or molecules is a pristine quantum system under this paradigm. Atoms have energy levels with transitions in the visible or ultraviolet range, which can be addressed with laser light. Due to these energy scales, experiments can be conducted in room temperature apparatus. Information about their state can be gathered using state-dependent fluorescence. Atoms are used to create highly accurate frequency references since they form identical standard systems. So, scaling of the quantum device can be achieved by adding more atoms, each of which will have identical properties. To achieve a Hamiltonian of interest, we need to engineer strong interactions in the system which is particularly advantageous to do with ions; since, as charged atoms, they interact strongly *via* the Coulomb force. A set of ions trapped in a common confining potential will exhibit a common set of motional modes. By coupling these modes to the internal (spin) degrees of freedom, we can engineer interactions between the spins, essentially using the motional modes as an information carrier or bus.[11] Additional details about the trapped ion approach can be found in Sec. II — Technical Addendum below.

52

## References

1 T. P. Harty, D. T. C. Allcock, C. J. Ballance, L. Guidoni, H. A. Janacek, N. M. Linke, D. N. Stacey, and D. M. Lucas, *High-fidelity preparation, gates, memory, and readout of a trapped-ion quantum bit*, Phys. Rev. Lett. **113**, 220501 Nov (2014).

2 K. Kim, M.-S. Chang, R. Islam, S. Korenblit, L.-M. Duan, and C. Monroe, *Entanglement and tunable spin-spin couplings between trapped ions using multiple transverse modes*, Phys. Rev. Lett. **103**, 12050 Sep (2009).

3 C. Schneider, D. Porras, and T. Schaetz, *Experimental quantum simulations of many-body physics with trapped ions*, Rep. Prog. Phys. **75**, no. 2, 024401 (2012).

4 R. Islam, C. Senko, W. C. Campbell, S. Korenbilt, J. Smith, A. Lee, E. E. Edwards, C.-C. J. Wang, J. K. Freericks, and C. Monroe, *Emergence and frustration of magnetism with variable-range interactions in a quantum simulator*, Science **340**, 583–587 May (2013).

5 C.-W. Chou, C. Kurz, D. B. Hume, P. N. Plessow, D. R. Leibrandt, and D. Leibfried, Preparation and coherent manipulation of pure quantum states of a single molecular ion, Nature **545**, 203–207 May (2017).

6 N. Huntemann, C. Sanner, B. Lipphardt, C. Tamm, and E. Peik, *Single ion atomic clock with $3 \times 10^{-18}$ systematic uncertainty*, Phys. Rev. Lett. **116**, 063001 Feb (2016).

7 S. Debnath, N. M. Linke, C. Figgatt, K. A. Landsman, K. Wright, and C. Monroe, *Demonstration of a small programmable quantum computer module using atomic qubits*, Nature **536**, 63–66 Aug (2016).

8 T. Monz, D. Nigg, E. A. Martinez, M. F. Brandl, P. Schindler, R. Rines, S. X. Wang, I. L. Chuang, and R. Blatt, *Realization of a scalable shor algorithm*, Science **351**, no. 6277, 1068–1070 (2016).

9 D. P. DiVincenzo, *The physical implementation of quantum computation*, Fortschritte der Physik **48**, no. 911, 771–783 (2000).

10 C. A. P´erez-Delgado and P. Kok, *Quantum computers: Definition and implementations*, Phys. Rev. A **83**, 012303 Jan (2011).

11 D. J. Griths, *Introduction to Electrodynamics*, Prentice Hall, 3rd ed. (1999).

### 4.3.4. *Via other approaches using quantum dots*

Ingmar Swart
*Utrecht University*

With respect to 2D quantum metamaterials, there currently are no reports that explicitly deal with electron-electron interactions in these systems. It should however be possible to include spin-orbit interactions, for example by building lattices from heavy atoms or by utilizing the surface state electrons of heavy metals, e.g., Re(0001).

4.3.5. *Via photons*

Garnett Bryant

*National Institute of Standards and Technology*

Quantum simulations, especially of the Hubbard model are done typically to determine ground state phase diagrams and to understand the competing effects of many-body interactions, single-particle hopping, disorder, and quantum fluctuations in simple models. [Note: strictly speaking "hopping" is a quantum fluctuation, at least in the Hubbard model.] Applications of dopant arrays to photonics require an understanding of many-body excited states and responses. Dopant arrays could be used both to do these excited state simulations and to realize the atom-based photonic structures being simulated. Dopant arrays could be used to simulate excited state dynamics, many-body entanglement, collision, and interference of many-body excitations and to implement atom-scale quantum and photonic applications. Exact simulations of many-body excitations can be done computationally for short chains. Such computations show how the excitation spectra can exhibit the ground state phase transitions, how quantized plasmons develop in the linear response regime, identify the bosonic character of these plasmonic states and determine when this quantization breaks down. However, understanding the transition to classical response requires simulations of much larger systems. Including spin or higher dimensions requires much larger systems. Understanding the dynamics of excitations, such as interference or collision between excitations requires even much larger systems. To address these fundamental issues, analog quantum simulations with large dopant arrays will be necessary.

Computational simulations have considered short, perfectly regular chains of identical atoms as a function of the interaction strength relative to the hopping energy. Quantum simulations with fabricated dopant arrays are needed to extend the results to longer chains, especially to explore the dynamics of many-body excitations. However, in such simulations the disorder in the dopant placement will be a critical issue. Much needs to be done to understand how sensitive results will be to disorder, especially when excitations are simulated. Development of improved precision of dopant placement would help to eliminate the effects of disorder and, at the same time, provide a way to develop controlled tests of disorder. Improved control of the dopant placement would also provide a way to tune the strength of interaction relative to hopping so that phase diagrams could be mapped. A range of new simulations with arrays could be done by using different types of dopant atoms and different arrangements or orderings of dopants. Developing this capability would greatly extend the type of simulations that could be done and could make assessable simulations of exotic materials with complex unit cells.[1]

## References

1    K. J. Morse, R. J. S. Abraham, A. DeAbreu, C. Bowness, T. Richards, H. Riemann, N. V. Abrosimov, P. Becker, H.-J. Pohl, M. L. W. Thewalt and S. Simmons, *A photonic platform for donor spin qubits in silicon*, Science Advances, e1700930 (2017).

## 4.4. Can quantum Monte Carlo methods (on classical computers) interface with the interpretation of experiments on precision-placed dopant atoms in semiconductors?

Richard Scalettar

*University of California, Davis*

Despite their limitations, several Quantum Monte Carlo approaches have a range of validity that enables a robust interface with proposed experiments using precision-placed dopant atoms in silicon.

The first is Determinant Quantum Monte Carlo (DQMC). This method provides an exact solution of the Hubbard Hamiltonian on lattices of several hundreds of sites (an order of magnitude larger than exact diagonalization) at temperatures down to about 1/40 of the electronic bandwidth $W$ for generic parameters. Because the method is formulated in real space, it is straightforward to incorporate general forms of the inter-site hopping and local repulsions and site energies. Thus, the effect of any randomness in dopant placement and size can be directly modeled.

One example of the potential interface between QMC and experiment is given in Fig. 4(a) and (b). Here we address whether randomness in the on-site interaction $U$ or in the inter-site hopping $t$ is more likely to affect the antiferromagnetic (AF) order, which occurs at half-filling. In Fig. 4(c), we also show how the nearest neighbor AF correlations can be affected in a single realization of randomness in the local interaction.

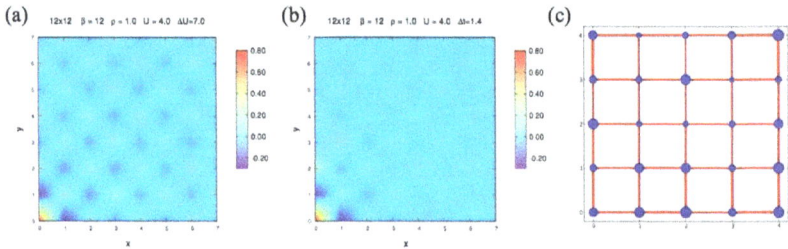

Fig. 4. (a) and (b) Antiferromagnetic correlations of the disordered Hubbard Hamiltonian on a $12 \times 12$ lattice at temperature $T = t/12 = W/96$. The density $\rho = 1$ (half-filling). (a) The onsite interactions vary uniformly in the range $U \in (0.5,7.5)$ about a median strength $U/t = 4$ which is one-half the non-interacting bandwidth. (b) The hopping varies with $t \in (0.3, 1.7)$. It is seen that randomness in the hopping leads to spin correlations that are considerably more 'washed out' than randomness in the interaction strength. Because periodic boundary conditions are used, correlations are shown only out to 7 sites away. (c) Simulation for a particular instance of disorder in $U$. The area of the circles is proportional to the strength of local $U$ on each site, the system is at half filling, the average $<U>$ is $8t$ and $\Delta U$ is $4t$. The thickness of the bonds is proportional to the simulated strength of the nearest-neighbor spin correlation.

For dopant array sizes currently accessible in experiments, exact diagonalization (ED) and the numerical linked cluster expansion method can also be employed to study inhomogeneous Hubbard models with customized local parameters. These approaches complement DQMC, especially in connecting to experiments, since they can directly access transport properties that are less straightforward to measure using QMC. ED has, for example, been used to study the evolution of dynamic quantities including the pseudo-gap and the spectral function (measured in angle-resolved photoemission) for disordered clusters.[1] An expanded discussion of this approach can be found in the Section II — Technical Addendum below.

## References

1    Simone Chiesa, Prabuddha B. Chakraborty, Warren E. Pickett, and Richard T. Scalettar, *Disorder-induced stabilization of the pseudogap in strongly correlated systems*, Phys. Rev. Lett. **101**, 086401 (2008).

## 5. Conclusions

James H. G. Owen[1] and Wiley P. Kirk[2,3]

[1]*Zyvex Labs,* [2]*University of Texas at Arlington,* [3]*3D Epitaxial Technologies*

### 5.1.1. *Overarching themes*

The workshop aimed to explore different approaches in the experimental realization of quantum metamaterials with properties that are tunable, so as to explore strong correlation physics, especially the Fermi–Hubbard model.

Major foci of the workshop concentrated on the concept of 2D quantum metamaterials from both experimental and theoretical perspectives, on approaches that lead to Analog Quantum Simulation (AQS), and whether a new embodiment of a 2D quantum metamaterial as determined by precisely placed dopants in semiconductors can also lead to AQS. Finally, it addressed the question does this new embodiment have any advantages or disadvantages over existing embodiments, such as cold atoms, trapped ions, *etc.*?

For cold atoms, the Hubbard model does not appear to be a model filled with approximations, but instead is a quantitatively accurate, rigorously derivable model for atoms in optical lattices in the ultra-cold regime. This makes it very useful to explore Hubbard physics, and even to add new factors into the system, such as magnetic fields in a controllable fashion. However, these systems are only stable at very low temperatures, and for the most interesting physics, even lower temperatures yet and currently unobtainable, may be required. Trapped ions can be operated at room temperature and show strong interactions *via* their Coulombic interactions. Each ion is essentially identical, and the level of disorder is low. They have strong coherence and are therefore a good candidate for use as qubits in quantum computers. However, the challenge is to scale up the achievable array sizes to the point where they would outstrip classical computers. Other approaches such as quantum dots have the strength of being achievable with well-known semiconductor processing techniques but suffer from imperfection and disorder.

As alluded to by Sven Rogge in a recent (2018) talk at a American Vacuum Society meeting in Long Beach, CA, the dopant-in-Si embodiment combines many of the advantages of other methods, with "cold-atom-like" mapping to spin $S = \frac{1}{2}$ models, "ion-trap like" precision control, and the use of a familiar semiconductor platform, Si, as is the case for quantum dot approaches. Moreover, it has some advantages, such as the ability to tune the spin-orbit coupling according to the choice and number of dopants at each array element (and even to switch to *p*-type dopants and explore hole-array behavior). Different lattice parameters and different lattices can be explored simply by changing the locations of the placed dopants; triangular, hexagonal, and even quasicrystal lattices are achievable, which would allow effects such as magnetic frustration to be explored. While there will inevitably be some disorder caused by fabrication processes, there is no theoretical limit to the size of the array that can be achieved with sufficiently accurate patterning; and the array, once fabricated, is air-stable for a long lifetime, with the dopant atoms trapped in fixed positions inside a silicon lattice.

### 5.1.2. *Suggested future work*

Future exploration in this direction should focus on expanding the range of possible implementations and on improving the precision of the fabrication process. While experimental approaches continue to be optimized and expanded, the theoretical approaches must keep up by answering questions such as the 'saggy couch' problem — *i.e.* how large do arrays need to be before they mimic infinite arrays?

Work continues to expand the available set of dopant elements beyond P. Arsenic and boron dopant placement have both been demonstrated; heavier dopants such as Bi with stronger spin-orbit coupling, and rare-earths such as Er to connect to photonic implementations would be useful. For each new element introduced, the surface chemistry for selective dopant placement must be determined in a manner as has been done for $PH_3$, and the encapsulation process adjusted accordingly. While industrial and academic efforts continue to improve pattern precision, other attempts to create large arrays ($32 \times 32$ or $100 \times 100$ pixels) still show significant disorder and further tool development will be required. On the synthesis side of investigation, which necessarily occurs at low-growth temperatures, the overgrowth of the dopant layers with silicon of sufficient epitaxial quality remains challenging. A move from UHV to XHV ($10^{-12}$ Torr or better) appears to be necessary to minimize contamination in the grown film. This will become more crucial if a move is made from 2D arrays to 3D, where multiple cycles of array formation and overgrowth will be required. It will also be important to determine the level of disorder and contamination that will be acceptable.

In summary, by bringing the foundational 2D cold-atom array approach into a solid-state environment, leveraging 15 years of STM-based Si:P fabrication, and making the fabrication of arbitrary 2D arrays of dopants available to any researcher with a good UHV STM system (and experience with gas handling and Si MBE), the dopant-in-Si approach should make Hubbard physics research far more accessible to a broader range of researchers.

# Section II — Technical Addendum
(Expanded state-of-the-art: Theoretical and experimental perspectives)

## 6. Theoretical landscapes and numerical-computational status

Richard Scalettar[1] and Ehsan Khatami[2]

[1]*University of California, Davis,* [2]*San Jose State University*

### Introduction

The Hubbard model offers one of the simplest ways to get insight into how the interactions between electrons can give rise to insulating, magnetic, and even novel superconducting effects in a solid. It was written down in the early 1960's and initially applied to understanding the behavior of the transition metal monoxides (FeO, NiO, CoO), compounds which are antiferromagnetic insulators, yet had been predicted to be metallic by methods that treat strong interactions less carefully.

Over the intervening years, the Hubbard model has been applied to the understanding of many systems, from 'heavy fermions' in the 1980's, to high temperature superconductors in the 1990's, to spin-liquid systems and, in multi-orbital variants, to iron-pnictide and topological materials. Indeed, it is an amazing feature of the model that, despite its simplicity, it exhibits behavior relevant to many of the subtlest and potentially technologically useful properties of solid-state physics.

The Hubbard model has been studied by the full range of analytic techniques developed by condensed matter theorists, from simple mean field approaches to field theoretic methods employing Feynman diagrams, expansions in the degeneracy of the number of 'flavors' (spin, orbital angular momentum), *etc.* It has also been extensively attacked with numerical methods like exact diagonalization and quantum Monte Carlo (QMC).[1-24]

The objective of this technical addendum is to introduce the Hubbard model and the quantities of interest to experiment, leaving, however, many of the specific details of engineered Si to the main body of the report. We focus on reviewing the different *computational* approaches. Discussions of analytic methods can be found in.[25-28] For completeness, we begin with a simple introduction of the model, and a discussion of its solution in the non-interacting limit, since this emphasizes the structure of the energy bands and density of states in different geometries which are important as a foundation to the correlation effects.

### The Hubbard Hamiltonian

The Hubbard Hamiltonian treats the regular array of nuclear positions in a solid as a *fixed* set of lattice sites; in the first of a number of approximations to real solids, it does not account for lattice vibrations. The atoms are further simplified so as to accommodate only a single orbital, which can hold either a spin up or a spin down electron, or both. Acknowledging the screening of the Coulomb potential, the electrons interact with a repulsion $U$ if they sit on the same site. A number of methods are capable of taking longer-range Coulomb interactions also into account, however such calculations are computationally very challenging. The kinetic energy consists of an expression which allows electrons to move from one site to its neighbors. The energy scale $J$ which governs this 'hopping' is determined by the overlap of two wavefunctions on the pair of atoms. Since wavefunctions die off exponentially, hopping often is allowed only between the near neighbor sites.

Defining $c_{j\sigma}^{\dagger}(c_{j\sigma})$ to be the operators that create (destroy) electrons of spin $\sigma$ on lattice site j, the Hubbard Hamiltonian is written as

$$\hat{H} = -J \sum_{\langle j,l \rangle \sigma} c_{j\sigma}^{\dagger} c_{l\sigma} + U \sum_{j} n_{j\uparrow}\, n_{lj\downarrow} - \mu \sum_{j} (n_{j\uparrow} + n_{j\downarrow}), \qquad (1)$$

where $\langle j,l \rangle$ denotes that sites j and l are nearest neighbors. The situation where the filling is one electron per site is referred to as 'half-filling' since the lattice contains half as many electrons as the maximum number (two per site). On a bipartite lattice, half-filling occurs at chemical potential $\mu = U/2$. Studies of the Hubbard model often initially focus on the half-filled case because it exhibits Mott insulating behavior and antiferromagnetic order.

The physics of the Hubbard model is encoded in various two-particle correlation functions, in particular the space and imaginary time dependent spin-spin correlations,

$$C_{ij}(\tau) = \langle S_{i+j}^{-}(\tau) S_{i}^{+}(0) \rangle \qquad\qquad S_{i}^{+} = e^{\tau \hat{H}} c_{i\uparrow}^{\dagger} c_{i\downarrow} e^{-\tau \hat{H}}, \qquad (2)$$

and analogous observables for the charge and pairing,

$$D_{ij}(\tau) = \langle \rho_{i+j}(\tau)\rho_{i}(0) \rangle - \langle \rho_{i+j}(0)\rangle\langle\rho_{i}(0)\rangle \quad \rho_{i}(\tau) = e^{\tau \hat{H}}(c_{i\uparrow}^{\dagger}c_{i\uparrow} + c_{i\downarrow}^{\dagger}c_{i\downarrow})e^{-\tau \hat{H}} \quad (3)$$

$$P_{ij}(\tau) = \langle \Delta_{i+j}(\tau)\Delta_{i}^{\dagger}(0) \rangle \qquad\qquad \Delta_{i}^{\dagger}(\tau) = e^{\tau \hat{H}} c_{i\uparrow}^{\dagger} c_{i\downarrow}^{\dagger} e^{-\tau \hat{H}}. \qquad (4)$$

On a translationally invariant lattice these correlation functions depend only on the difference i–j. In the presence of randomness, averaging over different disorder realizations restores translation invariance. These real space quantities can be summed to obtain the structure factors at momenta p. One often focuses on the uniform [p = (0,0)] and antiferromagnetic [p = $(\pi, \pi)$] values.

The correlation functions in imaginary time $\tau$ can be integrated to yield magnetic, charge, pairing, and transport susceptibilities. Dynamical properties like the spectral function $A(p, \omega)$ can be obtained through analytic continuation with the maximum entropy method[18] in cases where the methodology provides

Table 1. List of the correlation functions of interest in the
Hubbard model [Eqs. (2–6)] and their connection to experiments.

| Correlation | Notation | Operators | Experiment |
|---|---|---|---|
| Greens Function | $A(\mathbf{p},\omega)$ | $c_{\mathbf{p},\sigma}$ | Angle-Resolved Photoemission |
| Magnetic Structure factor | $\tilde{S}(\mathbf{p},\omega)$ | $S^+ = c^\dagger_{\mathbf{p},\uparrow} c_{-\mathbf{p},\downarrow}$ | Inelastic Neutron Scattering |
| Conductivity | $\Lambda(\mathbf{p},\omega)$ | $j_x = iJ(c^\dagger_{i+\hat{x}} c_i - c_i c^\dagger_{i+\hat{x}})$ | Optics and Transport |
| Four Spin correlations | $R(\omega)$ | $S^+_{\mathbf{p}} \cdot S^-_{-\mathbf{p}}$ | Raman Spectroscopy |

data as a function of imaginary time, or more directly with techniques like exact diagonalization. In the former case, for $A(\mathbf{p}, \omega)$, we compute the imaginary time dependent Green's function and then invert:

$$G(\mathbf{p}\tau) = \langle c_{\mathbf{p}}(\tau) c^\dagger_{\mathbf{p}}(0) \rangle \qquad G(\mathbf{p}\tau) = \int dw \frac{e^{-\omega\tau}}{e^{\beta\omega} + 1} A(\mathbf{p}, \omega). \qquad (5)$$

For the dynamical spin susceptibility,

$$S(\mathbf{p}, \tau) = -\int_{-\infty}^{+\infty} d\omega \frac{\operatorname{Im}\tilde{S}(\mathbf{p}, \omega) e^{-\tau\omega}}{1 - e^{-\beta\omega}} \qquad (6)$$

with analogous expressions for charge $\operatorname{Im}\Pi(\omega)$, pairing $\operatorname{Im}P(\mathbf{p}, \omega)$, and current $\Delta(\mathbf{p}, \omega)$. The relative sizes of the charge, spin, pairing, and current gaps can be used to distinguish between different types of insulating phases.[29]

A more direct procedure for dynamics is described later when details of the exact diagonalization approach are provided. These quantities directly connect to experimental probes of materials, emphasizing the need to measure dynamics. See Table 1.

**Limit of No Interactions**

In the case of translationally invariant lattices, the Hubbard Hamiltonian is solved analytically by defining creation and destruction operators in momentum space.

$$c^\dagger_{\mathbf{p}\sigma} = \frac{1}{\sqrt{N}} \sum_j e^{i\mathbf{p} \cdot j} c^\dagger_{j\sigma}. \qquad (7)$$

Depending on the lattice geometry, there are different dispersion relations which reflect the energy $\epsilon(\mathbf{p})$ associated with momentum $\mathbf{p}$. For example, in the studied case of a square lattice

$$\epsilon(\mathbf{p}) = -2J(\cos p_x + \cos p_y).$$

The dispersion relation determines the density of states $N(E)$, which counts the number of ways in which the system can have a given energy $E$.

$$N(E) = \frac{1}{N} \sum_{\mathbf{p}} \delta\left(E - \epsilon(\mathbf{p})\right).$$

The momentum integrated 'spectral function' Eq. (5) is the generalization of the density of states to the situation when interactions are turned on ($U \neq 0$) and is one of the central quantities characterizing the metal-insulator transition.

In many cases, the behavior of the density of states forms the basis of the simplest understanding of the physics of the Hubbard Hamiltonian. On the square lattice, for example, $N(E)$ has a van-Hove singularity (Fig. 5, left) at half-filling ($E = 0$) which plays a fundamental role in the critical value of $U$ for which magnetic order onsets. This singularity, along with Fermi surface nesting, was also suggested to have implications for high temperature (cuprate) super-conductivity. The density of states of other geometries is also the foundation for determining the properties of associated materials. As an example, the honeycomb lattice of graphene has a linearly vanishing $N(E)$ (see Fig. 5, right) associated with the Dirac cones of its dispersion relation. The 'Lieb lattice' which forms the basis of a more sophisticated description of the $CuO_2$ sheets of the cuprates hosts a 'flat band' and topologically localized states. In this way, even the simple lattice structure of the Hubbard Hamiltonian can reflect basic band structure of materials, and hence be an appropriate starting point for an understanding of the deeper effects of electron-electron correlations.

With that background, we now turn to methods that can address the effects of those interactions. When possible, we frame our discussion in the language of engineered silicon systems.

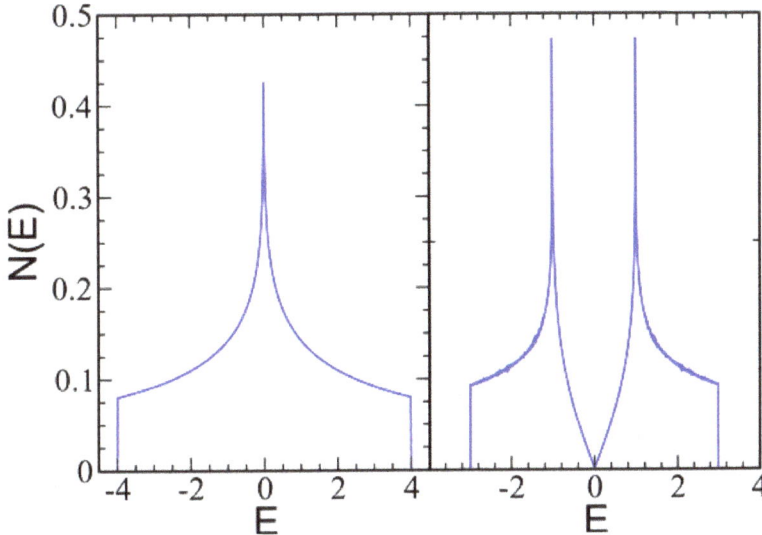

Fig. 5. The density of states, $N(E)$ of the square lattice Hubbard model (left) and the honeycomb lattice (right). Note the special features at half-filling ($E = 0$): the van-Hove singularity in the former, and the semi-metallic behavior in the latter.

**Exact Diagonalization**

For small enough lattices (in particular for small dopant arrays under consideration in this report), the method of choice for solving the Hubbard model is exact diagonalization (ED). In this approach, the Hamiltonian matrix is expressed in some basis (usually the many-body particle number basis for the Fermi–Hubbard model) and is then fully diagonalized. The access to the exact eigenvalues and eigenvectors of the system allows for the calculation of any static or time/frequency dependent quantity, including the transport properties. Certain symmetries of the Hamiltonian and conservation laws, such as the SU(2) symmetry, the conservation of total number of particles or spin, are used to block diagonalize the matrix and reduce the memory and computational time. The size of the Hilbert space, and hence, the size of the Hamiltonian matrix, grows exponentially with the size of the system and the diagonalization is simply not feasible beyond ~10 sites. For a 10-site system, the matrix corresponding to the sector with 10 particles and a total spin of zero has a dimension of 63,504, which would require about 120 GB of random-access memory (RAM) and a few tens of hours (depending on the processor) to be fully diagonalized using an optimized linear algebra package (LAPACK).

Observables like the charge correlation function of Eq. (2) can easily be expressed in terms of the energy eigenstates $|\psi_\alpha\rangle$ and eigenvalues $E_\alpha$:

$$C_{ij}(t) = \langle S_{i+j}^-(t)S_i^+(0)\rangle$$

$$= Z^{-1} \sum_\alpha e^{-\beta E_\alpha} \langle \psi_\alpha | S_{i+j}^-(t)S_i^+(0)|\psi_\alpha\rangle$$

$$= Z^{-1} \sum_\alpha e^{-\beta E_\alpha} \langle \psi_\alpha | e^{itH} S_{i+j}^-(0) e^{-itH} S_i^+(0)|\psi_\alpha\rangle$$

$$= Z^{-1} \sum_{\alpha,\beta} e^{-\beta E_\alpha} \langle \psi_\alpha | e^{-itH} S_{i+j}^-(0)|\psi_\beta\rangle \langle \psi_\beta | e^{itH} S_i^+(0)|\psi_\alpha\rangle$$

$$= Z^{-1} \sum_{\alpha,\beta} e^{-\beta E_\alpha} e^{it(E_\alpha - E_\beta)} \langle \psi_\beta | S_i^+(0)|\psi_\alpha\rangle \langle \psi_\alpha | S_{i+j}^+(0)|\psi_\beta\rangle, \qquad (8)$$

where $Z = \sum_\alpha e^{-\beta E_\alpha}$, is the partition function and $\beta = 1/k_B T$ is the inverse temperature. Eq. (8) is the real time analog of Eq. (2).

The appearance of $E_\alpha - E_\beta$ emphasizes connection of the expression for the correlation function to the excitation energy scales in the system. The other key components are the matrix elements of the operators to be measured. The strength of the exact diagonalization approach is the ability to access these quantities directly, without recourse to methods like analytic continuation.

For large diagonalizations such as the one mentioned above for a 10-site system, one may take advantage of multi-threaded features in Intel's math kernel library (MKL) in case of large number of processors per node, or distributed-memory linear algebra packages, such as ScaLAPACK, which can employ massively parallel environments to reduce the computational time, and get around memory issues by distributing the Hamiltonian and other large matrices between nodes during the calculation.

Full diagonalization scales as $O(N^3)$, where $N$ is the matrix dimension. The same scaling applies to the calculation of the dynamical correlation functions after the diagonalization step, often with a much larger prefactor. In those cases, parallel schemes, such as message passing interface (MPI) or OpenMP can be utilized to distribute the computation load.

Results for the Hubbard model on system sizes that can be exactly diagonalized are expected to vary significantly depending on the size and geometry of the clusters considered. Those changes are expected to be especially notable in the weak-coupling region where the strength of the interaction strength is smaller than the noninteracting bandwidth. However, this can be advantageous in comparing with experiments on dopant arrays since currently available systems are far from the thermodynamic limit and each can have a unique geometry and set of model parameters. Another advantage of exact diagonalization over most other numerical methods for the Hubbard model is that any geometry and virtually any variant of the model can be simulated. This is especially useful for dopant arrays since hopping amplitudes or Coulomb interactions are expected to extend up to a few dopant sites in range.

### The Lanczos Algorithm

Current experiments with dopant arrays on Si can be done at extremely low temperatures, one of their major potential advantages over optical lattices. In cases where only the ground state, or even a few excited states, are sufficient to estimate properties at experimentally relevant temperatures, one can take advantage of the Lanczos algorithm to go to larger system sizes, typically about a factor of two larger than exact diagonalization. The Lanczos algorithm offers an iterative method to diagonalize huge matrices,

$$H|\Phi_1\rangle = e_1|\Phi_1\rangle + b_2|\Phi_2\rangle$$
$$H|\Phi_n\rangle = e_n|\Phi_n\rangle + b_{n+1}|\Phi_{n+1}\rangle + b_n|\Phi_{n-1}\rangle \tag{9}$$

based on the recurrent procedure

$$e_n = \langle\Phi_n|H|\Phi_n\rangle$$
$$|\Phi_{n+1}\rangle = H|\Phi_n\rangle - e_n|\Phi_n\rangle - b_n|\Phi_{n-1}\rangle. \tag{10}$$

By keeping only a few arrays of size $N$ during the calculation, one can access a limited, but nevertheless highly useful portion of the spectrum — the ground state and low-lying excited states. Through the successive operation of the Hamiltonian on an initially random state in Eq. (9), the basis is transformed to the basis for a Krylov subspace in which the matrix is tridiagonal, much smaller in dimension, and with a spectrum that approaches that of the original matrix starting at the ground state as the number of iterations increases. A simple QR algorithm can then diagonalize the tridiagonal matrix to obtain the lowest eigenvalues.

There are well-known stability issues associated with the Lanczos algorithm, preventing one from continuing the iterative process to obtain the full spectrum. Often these issues are caused by the loss of orthogonalization between basis vectors of the new subspace, which can be overcome by keeping more vectors of size $N$ during the calculation and adding an increasingly expensive re-orthogonalization step as the number of iterations is increased. Nevertheless, the algorithm remains a powerful one for the low-temperature physics of quantum lattice models.

## Numerical Linked Cluster Expansions

In numerical linked-cluster expansions (NLCEs)[30-33] an extensive property of the model on a finite or infinite lattice is expressed as a sum over contributions from all the clusters that can be embedded in the lattice; $viz.$

$$P = \frac{1}{L} \sum_c W_p(c),  \tag{11}$$

where $P$ is the extensive property per site, $L$ is the system size (can be infinity), and $W_p(c)$ is the contribution to property $P$ from cluster $c$. The system is either in the thermodynamic limit or is larger than what can be treated exactly, for example using exact diagonalization. In the case of infinite lattice size, the factor $1/L$ can be removed if we consider the contributions only from those clusters that are not related through translational symmetry. If the model retains the underlying point-group symmetries of the lattice, the contributions from all the clusters that are related by point-group symmetry transformations can also be combined.

The contribution of each cluster $W_P(c)$ is calculated through the inclusion-exclusion principle and using the exact knowledge of the property on that cluster and smaller clusters via ED.

$$p(c) = \sum_{s \subseteq c} W_p(s) \qquad W_p(c) = p(c) - \sum_{s \subset c} W_p(s)  \tag{12}$$

where the sum runs over the cluster $c$ and all of its sub-clusters $s$. Starting from the smallest cluster in the expansion (typically a single site) for which $\sum_{s \subset c} W_p(s) = 0$, one can obtain $W_p(c)$ for larger clusters up to a certain size until

$p(c)$ can no longer be exactly known. Exact diagonalization is used to calculate $p(c)$.

NLCEs can enjoy many of the advantages of the ED, e.g., virtually any model can be simulated, and one has access to the full partition function and all the static and time dependent correlation functions. The series, in its region of convergence, also produces exact results with no systematic or statistical errors. That is especially useful when it is written for the lattice model in the thermodynamic limit. The main disadvantage can be the limitation in temperature. For models with divergent correlations in the ground state, the convergence of the series is lost below the temperature where the correlation length grows beyond the order of the largest clusters in the series. For the uniform Hubbard model in the thermodynamic limit with a repulsive interaction, the dominant correlations at low temperatures are antiferromagnetic with an exchange constant that is inversely proportional to the interaction $U$ in the strong-coupling regime. Therefore, the lowest convergence temperature in the NLCE decreases with increasing $U$ at half filling and can be lower than what quantum Monte Carlo methods can reliably access due to sampling issues that arise at large $U$.[34-36]

NLCEs can be utilized to calculate thermodynamic as well as transport properties (see discussion on transport properties below) of the dopant arrays in Si for sizes larger than those that can be solved using ED. Since there can be significant disorder in the model parameters, leading to an entanglement that will remain short ranged even at the lowest temperatures, it would not be unexpected to find that the NLCE with a finite number of terms converges in a wide range of temperatures, including the ground state.

## Determinant Quantum Monte Carlo

In the determinant QMC method,[1] a path integral for the partition function $Z$ is constructed by discretizing the inverse temperature $\beta = L\Delta\tau$, so that the full imaginary time evolution operator can be written as the product of $L$ terms. The small parameter $\Delta\tau$ allows for the 'Trotter approximation' to separate the exponential of the kinetic $\hat{K}$ (the terms including the hopping $J$ and chemical potential $\mu$) and potential $\hat{V}$ (involving the on-site Hubbard interaction $U$) energies:

$$Z = \mathrm{Tr}\left[e^{-\beta\hat{H}}\right] = \mathrm{Tr}\left[e^{-\Delta\tau\hat{H}} \cdots e^{-\Delta\tau\hat{H}}\right] \approx \mathrm{Tr}\left[e^{-\Delta\tau\hat{K}}e^{-\Delta\tau\hat{V}} \cdots e^{-\Delta\tau\hat{K}}e^{-\Delta\tau\hat{V}}\right]. \quad (13)$$

On every site $j$ and imaginary time slice $\tau$, the terms $e^{-\Delta\tau\hat{V}}$ are rewritten via the discrete Hubbard–Stratonovich transformation,[37]

$$e^{-U\Delta\tau n_\uparrow n_\downarrow} = \frac{1}{2}e^{-\frac{U\Delta\tau(n_\uparrow+n_\downarrow)}{2}}\sum_{S=\pm 1} e^{\lambda S(n_-\uparrow n_\downarrow)} \quad (14)$$

so that up and down fermions no longer couple to each other, instead they move in a space and imaginary time dependent auxiliary field $S(j,\tau)$. The coupling constant $\lambda$ obeys $\cosh \lambda = e^{\frac{U\Delta\tau}{2}}$. This transformation converts Eq. (13) into a trace over quadratic forms of the fermion operators, and allows them to be integrated out analytically, resulting in the product of the determinants of two matrices $M_\sigma$, one for each spin species. The partition function $Z = \mathrm{Tr} e^{-\beta H}$ is then a sum over all configurations of the Hubbard–Stratonovich field, which is sampled stochastically with both single spin flip moves[1] and 'global' updates which decrease autocorrelation time.[38]

DQMC allows the exact solution of tight binding Hamiltonians like Eq. (1) on finite spatial clusters, up to statistical errors generated by the Monte Carlo sampling. Typical simulation times are several hours for a lattice of 100 sites on a desktop computer. Systematic 'Trotter errors' from the discretization of $\beta$ can be extrapolated to zero.[39-41] Continuous time methods also exist, in related algorithms. These methods are advantageous since there is no Trotter error,[42-44] especially when quantities of very high precision such as energy and double occupancy are being measured. DQMC currently allows simulations of several hundred up to one thousand spatial sites, depending on the strength of $U$ and the inverse temperature,[45] and order of magnitude larger than exact diagonalization (see Fig. 6). The chief limitation of DQMC is the sign problem.[46-52] Constrained

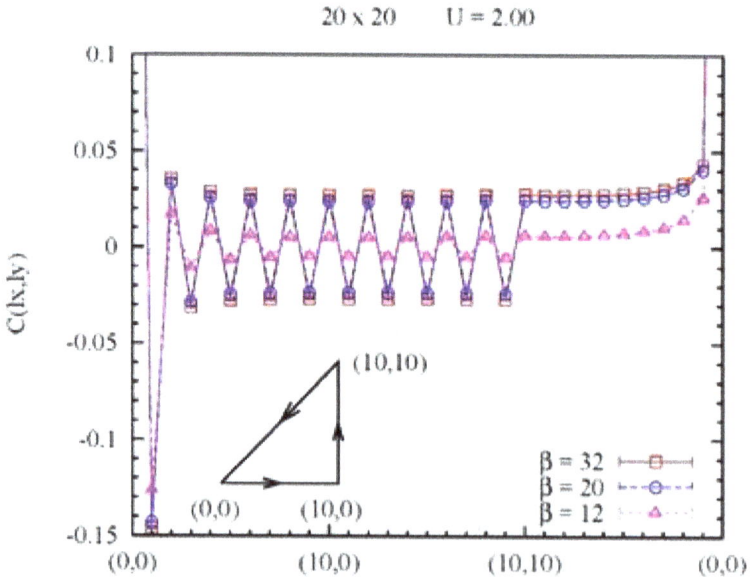

Fig. 6. DQMC results for the equal time magnetic correlations $C_{i,j}(\tau = 0) = \langle S^-_{i+j}(0)S^+_i(0)\rangle$ for $j = i + (l_x l_y)$ [see Eq. (2)]. Simulations were done on a $20 \times 20$ lattice at $U/t = 2$ and three inverse temperatures $\beta$. As $\beta$ increases, long range antiferromagnetic correlations develop.

path methods have been developed to address the sign problem in DQMC. However, they rely on an approximation in the form of assuming a particular structure for the nodes of the fermion wave function.[20]

The matrix inverses $G_{j\sigma} = M_{j\sigma}^{-1}$ are the single particle fermion Green's functions. Their matrix elements directly determine the density and kinetic energy. The double occupancy, and, indeed all the various spin, charge, and pairing correlations of Eqs. (2, 4) are obtained by averaging appropriate product of elements of $G_{\uparrow}G_{\downarrow}$.

## Dynamical Mean Field Theory and its Cluster Extensions

Conventional 'static' mean field theory (MFT) was one of the earliest methods used to solve the Hubbard Hamiltonian. The approach begins by decoupling the interaction term, $Un_{j\uparrow}n_{j\downarrow} \rightarrow U(n_{j\uparrow}\langle n_{j\downarrow}\rangle + \langle n_{j\uparrow}\rangle n_{j\downarrow} - \langle n_{j\uparrow}\rangle\langle n_{j\downarrow}\rangle)$. After making a specific ansatz for the averages $\langle n_{j\sigma}\rangle$, the resulting Hamiltonian is then quadratic in the fermion operators and can be diagonalized. The physics is determined by finding the expectation values which minimize the free energy, or, equivalently, a self-consistent computation of $\langle n_{j\sigma}\rangle$. Static MFT can be used to determine magnetic and charge order (and, indeed, provided some of the first indications of 'stripe' formation,[53]) but suffers some very serious limitations. It tends to grossly overestimate the tendency for ordered phases. Even more significantly, since static MFT reduces the problem to noninteracting electrons coupled to average densities, the resulting excitations have infinite lifetime.

This latter problem is rectified by the Dynamical Mean Field Theory (DMFT).[54-63] The basic idea of DMFT is to replace the full lattice problem of Eq. (1) with a single-site 'impurity' problem in which the local Green's function is determined self-consistently. The mean field with which the impurity couples is allowed to fluctuate in imaginary time so that DMFT better models the effects of the electron-electron interaction $U$.

The Anderson Impurity Model (AIM) onto which DMFT maps the Hubbard Hamiltonian is given by,

$$H = \sum_p \epsilon_p a_p^\dagger a_p + \sum_{p\sigma} V_{p\sigma} \left(a_{p\sigma}^\dagger c_\sigma + c_\sigma^\dagger a_{p\sigma}\right) + Un_\uparrow n_\downarrow - \mu(n_\uparrow + n_\downarrow), \quad (15)$$

which describes a single electron mode ($c_\sigma$) hybridized with a bath ($a_{p\sigma}$). This AIM can be solved via a variety of techniques including the numerical renormalization group, iterated perturbation theory, the non-crossing approximation, and continuous time QMC, the last of which being the most challenging computationally, but at the same time the least biased. These methods yield the impurity Greens function $G_{imp}(\tau) = hT\ c(\tau)c^\dagger(0)i$ (where $T$ is the imaginary time ordering operator) whose Fourier transform is,

$$G_{\text{imp}}(i\omega_n) = \sum_p \frac{1}{i\omega_n + \mu - \epsilon(p) - \Sigma(p, i\omega_n)}. \tag{16}$$

The DMFT approximation is the replacement of $\Sigma(p, i\omega_n)$ by $\Sigma(i\omega_n)$, *i.e.* ignoring the momentum dependence of the self-energy. The DMFT equations are solved self-consistently from a starting guess for $\Sigma(i\omega_n)$. DMFT can be shown to provide an exact solution of the Hubbard Hamiltonian in the limit of infinite dimension.[54,58]

Although we focus here on the use of DMFT to solve the Hubbard Hamiltonian, it is important to note that it has also revolutionized electronic structure calculations (the so-called 'LDA+DMFT' method[64]) since realistic band structures can be incorporated into $\in$ (p) in Eq. (16).

The underlying approximation of DMFT, that the self-energy is independent of momentum, can be systematically improved via extensions such as the Dynamic Cluster Approximation (DCA)[65,66] and Cluster DMFT.[67] Within the DCA, the self-energy is evaluated on a grid of $N_c$ momentum points. Since they build on the method described above, we will not review the details here. However, we show in Fig. 7 some state-of-the-art results for the *d*-wave superconducting transition in the 2D Hubbard model.[68]

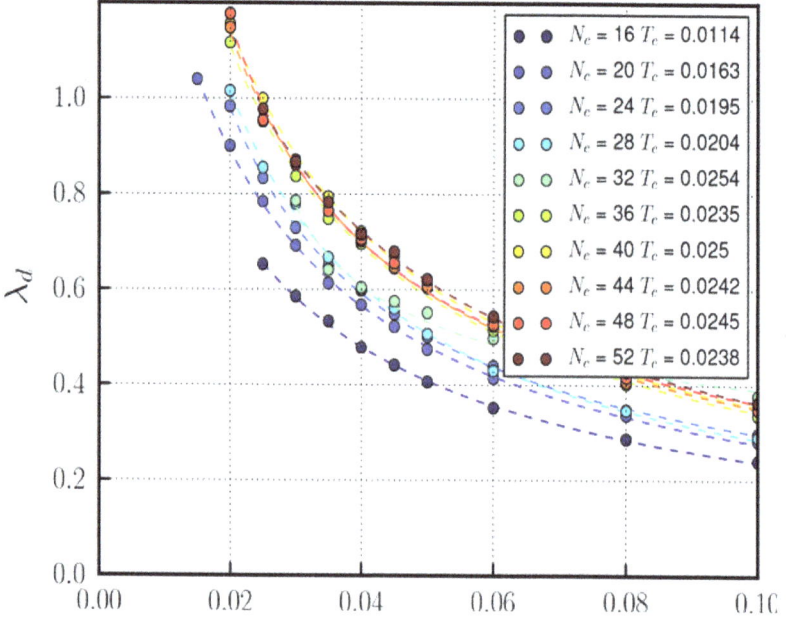

Fig. 7. The d-wave eigenvalue of the 2D Hubbard model for $U/t = 4$ and filling $\langle n \rangle = 0.9$. Data for different cluster sizes $N_c$ converge to $T_c/J \sim 0.024$, the temperature at which $\lambda_d \to 1$.

DMFT and its cluster extensions have several very significant advantages over real-space methods like DQMC. Specifically, they have a much better 'sign problem' (although this becomes less true as the cluster size increases) and they work directly in the thermodynamic limit. The latter fact enables them to extract transition temperatures via divergences of appropriate susceptibilities (Fig. 7) without having to make recourse to a laborious finite size scaling analysis as is required in, for example, DQMC.[45] For these reasons, much of our most reliable knowledge of the physics of the Hubbard Hamiltonian in the thermodynamic limit, including the question of the existence of a superconducting state with $d$-wave symmetry, has been through these approaches.

**Diagrammatic Quantum Monte Carlo**

Diagrammatic QMC [21,69-73] begins with a perturbative expansion of the partition function

$$Z = \mathrm{Tr}\, e^{-\beta H} = \mathrm{Tr}\, T\, e^{-\beta H_0} \exp\left[-\int_0^\beta d\tau H_1(\tau)\right]$$

$$= \sum_k (-1)^k \int_0^\beta d\tau_1 \cdots \int_{\tau_{k-1}}^\beta d\tau_k \mathrm{Tr}\left[e^{-\beta H_0} H_1(\tau_k) \cdots H_1(\tau_1)\right], \qquad (17)$$

where $H_0$ and $H_1$ are the Hubbard model kinetic and potential energy terms respectively. In contrast to DQMC, which samples a Hubbard–Stratonovich or phonon field with a weight given by fermion determinants, diagrammatic QMC instead samples the Feynman diagrams and integration variables (momentum-energy) with which they are constructed. The numbers and positions of the vertices are also sampled through insertion and removal of $\{\tau_l\}$, effectively sampling the variable $k$ in Eq. (17). For a given temperature and interaction strength, there is a peak in the distribution of the order of the diagrams contributing to Eq. (17), which allows for efficient sampling.

As with other numerical methods, significant improvements since its first introduction have been made to diagrammatic QMC, including the analytic summation of all connections to vertices, which reduces the phase space to be explored stochastically and also reduces or even eliminates, in some cases, the sign problem. The replacement of bare interaction vertices by exact two-body scattering amplitudes likewise is a way to perform parts of the diagrammatic sum analytically.

Applications of diagrammatic QMC have been made to the Hubbard model and also to electron-phonon systems [72] and frustrated spin systems.[73] Like DMFT, diagrammatic QMC can be combined with electronic structure theory[74,75] to provide more accurate modeling of real materials. (see Fig. 8)

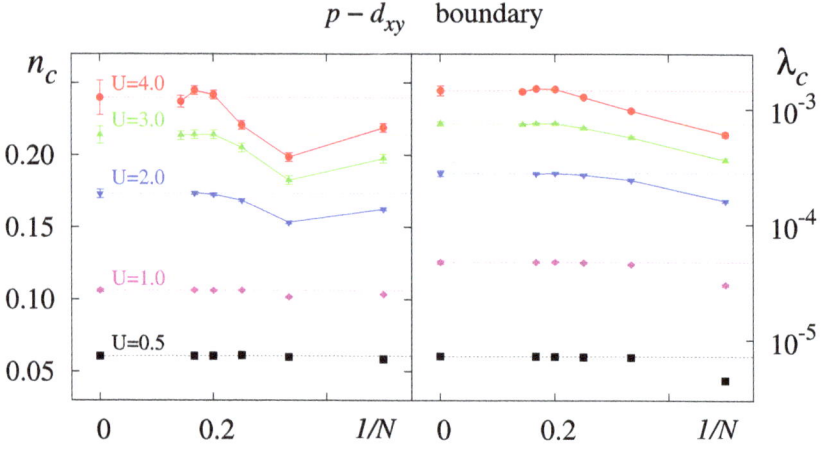

$p - d_{xy}$ boundary

Fig. 8. Some diagrammatic QMC results for the square lattice Hubbard model at densities $\langle n \rangle < 0.7$. Critical density (left) and critical coupling constant (right) for the phase boundary between $p$-wave and $d$-wave superfluid phases as a function of the expansion order $N$.

## Transport Properties

Spin and charge transport can be studied using the numerical methods discussed in this report. These properties will have the most relevance to initial dopant array experiments.

To study transport properties numerically, we look at the system's response to an external electric or magnetic field that is in general a function of both space and frequency. We calculate the alternating current (AC) in the linear response regime through Kubo's formula, which relates this current to the correlation functions in the unperturbed system.[76,77] For example, in the absence of any coherent response (Drude weight), the AC conductivity can be written as

$$\sigma_{xx}(q, \omega) = \frac{i}{\omega} \Lambda_{xx}(q, \omega) \,, \tag{18}$$

where $\omega$ is the frequency, and $\Lambda_{xx}(q, \omega)$ is the Fourier transform of the retarded correlation function of the current operator

$$\Lambda_{xx}(q, t - t') = -i\Theta(t - t')\langle [j_x(-q, t), j_x(q, t')] \rangle \,. \tag{19}$$

In which $t$ and $t'$ are time, $\Theta$ is the step function, and $j_x(q, t) = e^{itH} j_x(q) e^{-itH}$ is the time-dependent current operator at wavevector q (we have taken $\sim = 1$). The charge current operator can be obtained via the continuity equation and takes the following form for the Hubbard model:

$$j_x(q) = iJ \sum_{l,\sigma} e^{iq \cdot r_l} \left( c^\dagger_{l+x\sigma} c_{l\sigma} - c^\dagger_{l\sigma} c_{l+x\sigma} \right) \,, \tag{20}$$

where $J$ is again the hopping amplitude. Taking the uniform limit $q = 0$, the real part of the AC conductivity, which is a quantity measured in the engineered Si experiments, can be simplified to

$$\text{Re } \sigma_{xx}(\omega) = -\frac{2}{\omega} \text{Im} \int_0^\infty dt \, e^{i\omega t} \text{Im}\langle j_x(t)j_x(0)\rangle. \tag{21}$$

Other equivalent expressions for $\text{Re } \sigma(\omega)$ that use only the real part or both the real and imaginary parts of the current-current correlation function can be derived as well. For example, one can show that [76,78,79]

$$\text{Re } \sigma_{xx}(\omega) = \frac{(1 - e^{-\beta\omega})}{\omega} \text{Re} \int_0^\infty dt \, e^{i\omega t} \langle j_x(t)j_x(0)\rangle, \tag{22}$$

whose direct current (DC) limit ($\omega \to 0$), takes a simple form of

$$\text{Re } \sigma_{xx}^{DC} = \beta \text{ Re} \int_0^\infty dt \, \langle j_x(t)j_x(0)\rangle. \tag{23}$$

In principle, the time-dependent current correlator can be calculated in the ED or the NLCE in order to obtain the conductivity.[79] However, in the ED for small clusters, the correlator is expected to exhibit significant fluctuations at all times due to the boundaries, and therefore, any Fourier transform of the finite-time correlation function can lead to uncertainties in the conductivity. For this reason, it may be advantageous to take the Fourier transform first by explicitly expressing the time dependence of the current correlator.[76] The resulting formula takes the following form, often used in the ED study of the conductivity:[80]

$$\text{Re } \sigma_{xx}(\omega) = \pi \frac{(1 - e^{-\beta\omega})}{\omega Z} \times \sum_{n,m} e^{-\beta E_n} |\langle n|j_x(0)|m\rangle|^2 \delta(\omega + E_n - E_m), \tag{24}$$

where $Z$ is the partition function, $E_n$ is the eigenenergy of the $n^{\text{th}}$ eigenstate of the Hamiltonian, and $\delta$ is the delta function. This expression is of course directly analogous to Eq. (22).

In the NLCE, we can work with real time current correlation functions.[79] Even though the convergence of the series at a given temperature can now be lost beyond some maximal time, in the absence of boundary effects for the system in the thermodynamic limit, the fluctuations in time are expected to die off eventually and a Fourier transform of the Green's function in the converged region can still provide a good estimate for the AC conductivity, especially away from the zero frequency limit.

In Fig. 9, we show the AC *spin* conductivity for the Hubbard model in the thermodynamic limit. The spin current is defined similar to the charge current in Eq. (20) and is obtained by including the spin $\sigma/2$ as a multiplying factor inside the sum.[79]

(a)　　　　　　　　　　(b)　　　　　　　　　(c)

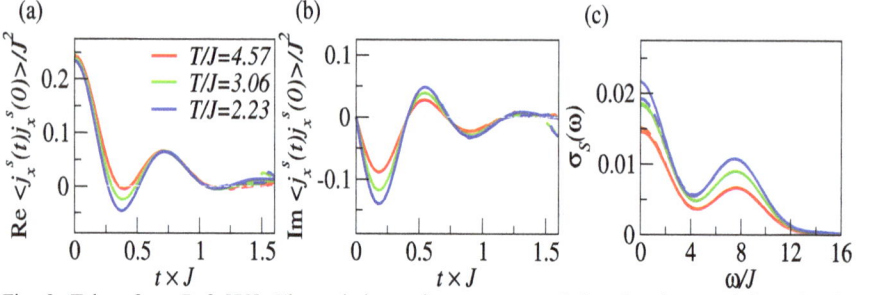

Fig. 9. Taken from Ref. [79]. The real time spin current correlation functions and the AC spin conductivity of the Hubbard model are shown for $U/J = 8$ at half filling. The AC spin conductivity is obtained from Eq. (22) (solid lines) and Eq. (21) (dashed lines) at the various temperatures, using the NLCE.

Even though the real time dependent quantities are not accessible in equilibrium quantum Monte Carlo methods, such as DQMC, DMFT, DCA, and diagrammatic QMC described earlier, the retarded current Green's function of Eq. (19) can be calculated on the imaginary time axis:

$$\Lambda_{xx}(q, \tau) = -\langle T j_{xx}(q, \tau) j_{xx}(q, 0) \rangle. \tag{25}$$

The AC conductivity can be obtained by Fourier transforming the imaginary-time Green's function to the Matsubara frequency space and then analytically continuing the imaginary frequency quantity to the real frequency axis $(i\omega_n \rightarrow \omega + \delta)$, a process that is numerically ill-defined, but can be accomplished using, e.g., the maximum entropy method. It amounts to taking the inverse of an integral such as an analog of Eq. (6),[81] viz.

$$\Lambda_{xx}(\tau) = \int\limits_{-\infty}^{\infty} \frac{d\omega}{\pi} \frac{-e^{-\tau\omega}}{1 - e^{-\beta\omega}} \operatorname{Im} \Lambda_{xx}(\omega). \tag{26}$$

An approximate form for the DC conductivity can be worked out for temperatures much lower than a characteristic frequency scale in the system below where the AC conductivity is constant. Starting from Eq. (26), it can be shown that in this regime

$$\sigma_{xx}^{DC} \approx \frac{\beta^2}{\pi} \Lambda_{xx}\left(\frac{\beta}{2}\right), \tag{27}$$

a form that can be used as a proxy for the actual DC conductivity of the system. Similar expressions have been used for the electronic spectral function, spin susceptibility, as well as conductivity.[81-83]

Another more robust proxy can be derived for the same temperature regime by incorporating information from the curvature of the current Green's function.[84,85]

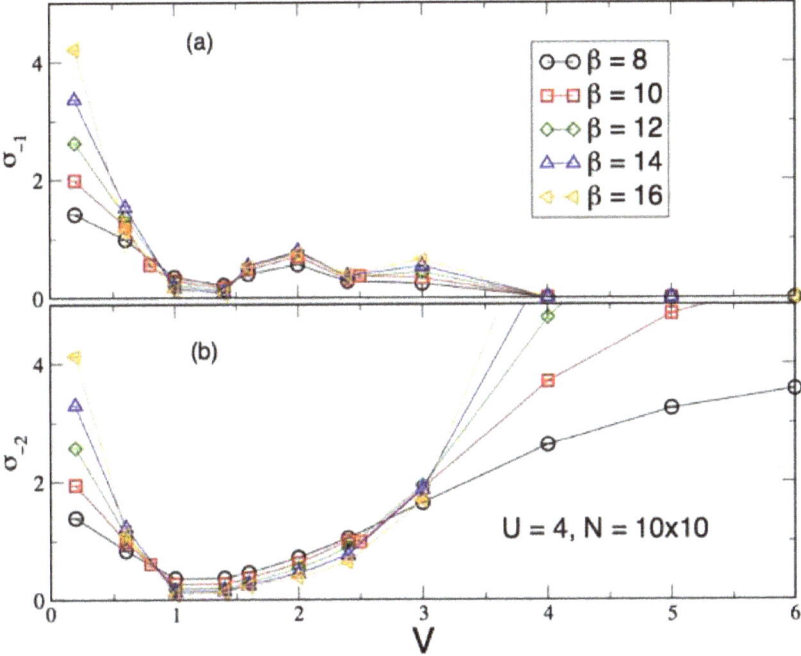

Fig. 10. An illustration of DQMC capabilities for obtaining conductivities via analytic continuation. The system being studied is an interface between a Mott insulator and a metal. Simulations were done on $10 \times 10$ layers at $U/J = 4$ and five inverse temperatures $\beta$.

Fig. 10 shows some DQMC results for the transport in a system with several weakly interacting (metallic) layers, connected to a Mott insulator via an interfacial hopping $V$. $\sigma_{-1}$ is the conductivity [obtained via Eq. (27)] in the metallic layer most proximate to the insulator. $\sigma_{-2}$ is the conductivity in the next deepest metallic layer. For $V$ larger than four times the intralayer hopping $J$, the conductivity $\sigma_{-1}$ right at the boundary vanishes as a consequence of the formation of magnetic singlets across the boundary. The conductivity deeper within the metal ($\sigma_{-2}$) then recovers as,

$$\sigma_{xx}^{DC} \approx \frac{2\pi \Lambda_{xx} \left(\frac{\beta}{2}\right)^2}{\Lambda_{xx}'' \left(\frac{\beta}{2}\right)}.  \tag{28}$$

## Conclusions

Here we have very briefly reviewed some of the technical details of the powerful set of computational approaches developed to solve the Hubbard Hamiltonian over the last three decades. These methods have given tremendous insight into the physics of the model. Unfortunately, one lesson has been that the Hubbard Hamiltonian supports a diverse set of possible low temperature phases, and that these often have free energies that differ by rather small amounts. As a consequence, there is the ongoing concern that innocuous approximations or finite size effects might alter the conclusions. This observation has driven more and more accurate algorithms, and also attempts such as optical lattice emulation and, potentially, the engineered materials approach of this report. The objective is a careful comparison of theory and experiment that will conclusively determine the strong correlation physics of the Hubbard model.

Of the methods described there, exact diagonalization, Lanczos, the Numerical Linked Cluster Expansion, and Determinant Quantum Monte Carlo all work directly in real space and have the capability of modeling uncertainties in atomic placement and on-site interaction strength. In addition, they have complementary capabilities in terms of lattice sizes, accessible energy and temperature scales, and ability to draw out transport properties. They therefore are likely to partner well with the first generation of engineered materials experiments. Once the initial testing and comparisons are done, one would expect the full panoply of techniques to be crucial.

**References**

1   R. Blankenbecler, D. J. Scalapino, and R. L. Sugar, *Monte Carlo calculations of coupled boson-fermion systems. I*, Phys. Rev. D **24**, 2278 (1981).

2   S. Sorella, S. Baroni, R. Car, and M. Parinello, *A Novel Technique for the Simulation of Interacting Fermion Systems,* Europhys. Lett. **8**, 663 (1989).

3   N. Prokof'ev, B. Svistunov, and I. Tupitsyn, *'Worm' algorithm in quantum Monte Carlo simulations*, Phys. Lett. A **238**, 253 (1998).

4   H. G. Evertz, G. Lana, and M. Marcu, *Cluster algorithm for vertex models*, Phys. Rev. Lett. **70**, 875 (1993).

5   H. G. Evertz, *The Loop Algorithm*, Advances in Physics **52**, 1 (2003).

6   O. F. Syljuasen and A. W. Sandvik, *Quantum Monte Carlo with directed loops*, Phys. Rev. E **66**, 046701 (2002).

7   F. Alet, S. Wessel, and M. Troyer, *Generalized directed loop method for quantum Monte Carlo simulations*, Phys. Rev. E **71**, 036706 (2005).

8   V. G. Rousseau, *Stochastic Green function algorithm*, Phys. Rev. E **77**, 056705 (2008).

9   K. Van Houcke, S. M. A. Rombouts, and L. Pollet, *Quantum Monte Carlo simulation in the canonical ensemble at finite temperature*, Phys. Rev. E **73**, 056703 (2006).

10  S. M. A. Rombouts, K. Van Houcke, and L. Pollet, *Loop Updates for Quantum Monte Carlo Simulations in the Canonical Ensemble*, Phys. Rev. Lett. **96**, 180603 (2006).

11    N. Kawashima, J. E. Gubernatis, and H. G. Evertz, *Loop algorithms for quantum simulations of fermion models on lattices,* Phys. Rev. B **50**, 136 (1994).

12    B. B. Beard and U. J. Wiese, Simulations of Discrete Quantum Systems in Continuous Euclidean Time, Phys. Rev. Lett. **77**, 5130 (1996).

13    N. V. Prokof'ev, B. V. Svistunov, and I. S. Tupitsyn, *Exact quantum Monte Carlo process for the statistics of discrete systems,* J. Expt. and Theor. Phys. **64**, 911 (1996).

14    N. V. Prokofev, B. V. Svistunov, and I. S. Tupitsyn, Exact, complete, and universal continuous-time worldline Monte Carlo approach to the statistics of discrete quantum systems, J. Expt. and Theor. Phys. **87**, 310 (1998).

15    H. Rieger and N. Kawashima, Application of a continuous time cluster algorithm to the two-dimensional random quantum Ising ferromagnet, Eur. Phys. J. B **9**, 233 (1999).

16    M. Jarrell, T. Maier, C. Huscroft, and S. Moukouri, A Quantum Monte Carlo algorithm for non-local corrections to the Dynamical MeanField Approximation,Phys. Rev. B **64**, 195130 (2001).

17    J. E. Gubernatis, M. Jarrell, R. N. Silver, and D. S. Sivia, *Quantum Monte Carlo simulations and maximum entropy: Dynamics from imaginary time data*, Phys. Rev. B **44**, 6011 (1991).

18    M. Jarrell and J. E. Gubernatis, Bayesian Inference and the Analytic Continuation of Imaginary-Time Quantum Monte Carlo Data, Phys. Rep. **269**, 133 (1996).

19    A. W. Sandvik, Stochastic series expansion method with operator-loop update, Phys. Rev. B **59**, R14157 (1999).

20    S. Zhang, J. Carlson, and J. E. Gubernatis, *Constrained path Monte Carlo method for fermion ground states*, Phys. Rev. B **55**, 7464 (1997).

21    E. Kozik, K. Van Houcke, E. Gull, L. Pollet, N. Prokof'ev, B. Svistunov and M. Troyer, *Diagrammatic Monte Carlo for correlated fermions*, Europhys. Lett. **90**, 10004 (2010).

22    P. K. V. V. Nukala, T. A. Maier, M. S. Summers, G. Alvarez, and T. C. Schulthess, *Fast update algorithm for the quantum Monte Carlo simulation of the Hubbard model*, Phys. Rev. B **80**, 195111 (2009).

23    E. Gull, P. Staar, S. Fuchs, P. K. V. V. Nukala, M. S. Summers, T. Pruschke, T. C. Schulthess, and T. A. Maier, *Submatrix updates for the continuous-time auxiliary-field algorithm*, Phys. Rev. B **83**, 075112 (2011).

24    E. Kozik, E. Burovski, V. W. Scarola, and M. Troyer, *N´eel temperature and thermodynamics of the half-filled three-dimensional Hubbard model by diagrammatic determinant Monte Carlo*, Phys. Rev. B **87**, 205102 (2013).

25    M. Rasetti, *The Hubbard Model- Recent Results*, World Scientific (1991).

26    Arianna Montorsi (ed), *The Hubbard Model*, World Scientific (1992).

27  F. Gebhard, The Mott Metal-Insulator Transition, Models and Methods, Springer (1997).

28  P. Fazekas, Lecture Notes on Electron Correlation and Magnetism, World Scientific (1999).

29  M. Vekic, J. W. Cannon, D. J. Scalapino, R. T. Scalettar, and R. L. Sugar, *Competition Between Antiferromagnetic Order and Spin Liquid Behavior in the Two-Dimensional Periodic Anderson Model at Half-Filling*, Phys. Rev. Lett. **74**, 2367 (1995).

30  M. Rigol, T. Bryant, and R. R. P. Singh, *Numerical linked-cluster approach to quantum lattice models*, Phys. Rev. Lett. **97**, 187202 (2006).

31  M. Rigol, T. Bryant, and R. R. P. Singh, *Numerical linked-cluster algorithms. i. spin systems on square, triangular, and kagom'e lattices*, Phys. Rev. E **75**, 061118 (2007).

32  M. Rigol, T. Bryant, and R. R. P. Singh, *Numerical linked-cluster algorithms. ii. $t - j$ models on the square lattice*, Phys. Rev. E **75**, 061119, (2007).

33  B. Tang, T. Paiva, E. Khatami, and Rigol, *Finite-temperature properties of strongly correlated fermions in the honeycomb lattice*, Phys., Rev. B **88**, 125127 (2013).

34  E. Khatami and M. Rigol, *Thermodynamics of strongly interacting fermions in two-dimensional optical lattices*, Phys. Rev. A **84**, 053611 (2011).

35  E. Khatami and M. Rigol, *Effect of particle statistics in strongly correlated two-dimensional hubbard models*, Phys. Rev. A **86**, 023633 (2012).

36  B. Tang, T. Paiva, E. Khatami, and M. Rigol, *Short-range correlations and cooling of ultracold fermions in the honeycomb lattice*, Phys. Rev. Lett. **109**, 205301 (2012).

37  J. E. Hirsch, *Discrete Hubbard-Stratonovich transformation for fermion lattice models*, Phys. Rev. B **28**, 4059 (1983).

38  R. T. Scalettar, R. M. Noack, and R. R. P. Singh, *Ergodicity at large couplings with the determinant Monte Carlo algorithm*, Phys. Rev. B **44**, 10502 (1991).

39  H. F. Trotter, *On the product of semi-groups of operators*, Proc. Amer. Math. Soc. **10**, 545 (1959).

40  M. Suzuki, *Generalized Trotter's formula and systematic approximants of exponential operators and inner derivations with applications to many-body problems*, Comm. Math. Phys. **51**, 183 (1976).

41  R. M. Fye, *New results on Trotter-like approximations*, Phys. Rev. B **33**, 6271 (1986).

42  A. N. Rubtsov and A. I. Lichtenstein, *Continuous time quantum Monte Carlo method for fermions: beyond auxiliary field framework*, Pis'ma JETP **80**, 67 (2004).

43   A. N. Rubtsov, V. V. Savkin, and A. I. Lichtenstein, *Continuous-time quantum Monte Carlo method for fermions*, Phys. Rev. B **72**, 035122 (2005).

44   S. Fuchs, E. Gull, L. Pollet, E. Burovski, E. Kozik, T. Pruschke, and M. Troyer, *Thermodynamics of the 3D Hubbard Model on Approaching the Neel Transition*, Phys. Rev. Lett. **106**, 030401 (2011).

45   C. N. Varney, C. R. Lee, Z. J. Bai, S. Chiesa, M. Jarrell, and R. T. Scalettar, *Quantum Monte Carlo Study of the 2D Fermion Hubbard Model at Half-Filling*, Phys. Rev. B **80**, 075116 (2009).

46   E. Y. Loh, J. E. Gubernatis, R. T. Scalettar, S. R. White, D. J. Scalapino, and R. L. Sugar, *The Sign Problem in the Numerical Simulation of Many Electron Systems*, Phys. Rev. B **41**, 9301 (1990).

47   N. Hatano and M. Suzuki, *Representation Basis in Quantum Monte Carlo Calculations and the Negative-Sign Problem*, Phys. Lett. A **163**, 246 (1992).

48   T. Nakamura, *Vanishing of the negative-sign problem of quantum Monte Carlo simulations in one dimensional frustrated spin systems*, Phys. Rev. B **57**, R3197 (1998).

49   S. Chandrasekharan and U.-J. Wiese, *Meron-Cluster Solution of Fermion Sign Problems*, Phys. Rev. Lett. **83**, 3116 (1999).

50   M. Troyer and U.-J. Wiese, *Computational complexity and fundamental limitations to fermionic quantum Monte Carlo simulations*, Phys. Rev. Lett. **94**, 170201 (2005).

51   J. Braun, J-W. Chen, J. Deng, J. E. Drut, B. Friman, C-T. Ma, and Y-D. Tsai, *Imaginary polarization as a way to surmount the sign problem in ab initio calculations of spin-imbalanced Fermi gases*, Phys. Rev. Lett. **110**, 130404 (2013).

52   V. I. Iglovikov, E. Khatami, and R. T. Scalettar, *Geometry dependence of the sign problem in quantum Monte Carlo simulations*, Phys. Rev. B **92**, 045110 (2015).

53   J. Zaanen and O. Gunnarsson, *Charged magnetic domain lines and the magnetism of high-Tc oxides*, Phys. Rev. B **40**, 7391 (1989).

54   M. Jarrell, *Hubbard model in infinite dimensions: A quantum Monte Carlo study*, Phys. Rev. Lett. **69**, 168 (1992).

55   Th. Pruschke, D. L. Cox, and M. Jarrell, *The Hubbard Model at Infinite Dimensions: Thermodynamic and Transport Properties*, Phys. Rev. B **47**, 3553 (1993).

56   A. Georges, G. Kotliar, W. Krauth, and M. Rozenberg, *Dynamical mean-field theory of strongly correlated fermion systems and the limit of infinite dimensions*, Rev. Mod. Phys **68**, 13 (1996).

57   A. Georges and G. Kotliar, *Hubbard model in infinite dimensions*, Phys. Rev. B **45**, 6479 (1992).

58   W. Metzner and D. Vollhardt, *Correlated Lattice Fermions in $d = \infty$ Dimensions*, Phys. Rev. Lett. **62**, 324 (1989)

59  G. Kotliar, S. Y. Savrasov, K. Haule; V. S. Oudovenko, O. Parcollet, and C. A. Marianetti, *Electronic structure calculations with dynamical mean-field theory*, Rev. Mod. Phys. **78**, 865 (2006).

60  D. Vollhardt, *Dynamical mean-field theory for correlated electrons*, Ann. Phys. **524**, 1 (2012).

61  A. Georges, *Strongly Correlated Electron Materials: Dynamical Mean-Field Theory and Electronic Structure*, AIP Conference Proceedings. American Institute of Physics Conference. Lectures on the Physics of Highly Correlated Electron Systems VIII, **715**, 3 (2004).

62  K. Held, *Electronic Structure Calculations using Dynamical Mean Field Theory*, Adv. Phys. **56**, 829 (2007).

63  A. Toschi, A. Katanin, and K. Held, *Dynamical vertex approximation: A step beyond dynamical mean-field theory*, Phys. Rev. B **75**, 045118 (2007).

64  K. Held, I. A. Nekrasov, G. Keller, V. Eyert, N. Blümer, A. K. McMahan, R. T. Scalettar, Th. Pruschke, V. I. Anisimov, and D. Vollhardt, *Realistic investigations of correlated electron systems with LDA+DMFT*, Psi-k Newsletter **56**, 65 (2003).

65  M. H. Hettler, M. Mukherjee, M. Jarrell, and H. R. Krishnamurthy, *Dynamical cluster approximation: Nonlocal dynamics of correlated electron systems*, Phys. Rev. B **61**, 12739 (2000).

66  T. Maier, M. Jarrell, T. Pruschke, and M. H. Hettler, *Quantum cluster theories*, Rev. Mod. Phys. **77**, 1027 (2005).

67  H. Park, K. Haule, and G. Kotliar, *Cluster Dynamical Mean Field Theory of the Mott Transition*, Phys. Rev. Lett. **101**, 186403 (2008).

68  P. Staar, T. Maier, and T. C. Schulthess, *Two-particle correlations in a dynamic cluster approximation with continuous momentum dependence: Superconductivity in the two-dimensional Hubbard model*, Phys. Rev. B **89**, 195133 (2014).

69  K. Van Houcke, F. Werner, E. Kozik, N. Prokof'ev, B. Svistunov, M. Ku, A. Sommer, L. Cheuk, A. Schirotzek, and M. Zwierlein, *Feynman diagrams versus Fermi-gas Feynman emulator*, Nature Physics **8**, 366 (2012).

70  Y. Deng, E. Kozik, N. V. Prokof'ev, and B. V. Svistunov, *Emergent BCS regime of the two-dimensional fermionic Hubbard model: ground-state phase diagram*, Europhys. Lett. **110**, 57001 (2015).

71  J Gukelberger, E. Kozik, L. Pollet, N. Prokof'ev, M. Sigrist, B. Svistunov, and M. Troyer, *p-wave Superfluidity by Spin-Nematic Fermi Surface Deformation*, Phys. Rev. Lett. **113**, 195301 (2014).

72  A. S. Mishchenko, N. Nagaosa, and N. Prokof'ev, *Diagrammatic Monte Carlo method for many-polaron problems*, Phys. Rev. Lett. **113**, 166402 (2014).

73  S. A. Kulagin, N. Prokof'ev, O. A. Starykh, B. Svistunov, and C. N. Varney, *Bold Diagrammatic Monte Carlo Method Applied to Fermionized Frustrated Spins*, Phys. Rev. Lett. **110**, 070601 (2013).

74  Andrea Marini, S. Ponc´e, and X. Gonze, *Many-body perturbation theory approach to the electron-phonon interaction with density-functional theory as a starting point*, Phys. Rev. B **91**, 224310 (2015).

75  K. Van Houcke, I. S Tupitsyn, and N. V. Prokof´ev, *Diagrammatic Monte Carlo and GW Approximation for Jellium and Hydrogen Chain*, Handbook of Materials Modeling: Methods: Theory and Modeling **1**, (2018).

76  G. D. Mahan, *Many-Particle Physics*. Plenum Press (1981).

77  D. J. Scalapino, S. R. White, and S. Zhang, *Insulator, metal, or superconductor: The criteria*, Phys. Rev. B **47**, 7995 (1993).

78  C. Karrasch, D. M. Kennes, and J. E. Moore, *Transport properties of the one-dimensional Hubbard model at finite temperature*, Phys. Rev. B **90**, 155104 (2014).

79  M. A. Nichols, L. W. Cheuk, M. Okan, T. R. Hartke, E. Mendez, T. Senthil, E. Khatami, H. Zhang, and M. W. Zwierlein, *Spin transport in a Mott insulator of ultracold fermions*, arXiv:1802.10018 (2018).

80  J. A. Riera and E. Dagotto, *Optical conductivity of the Hubbard model at finite temperature*, Phys. Rev. B **50**, 452 (1994).

81  N. Trivedi and M. Randeria, *Deviations from fermi-liquid behavior above $T_c$ in 2d short coherence length superconductors*, Phys. Rev. Lett. **75**, 312 (1995).

82  M. Randeria, N. Trivedi, A. Moreo, and R. T. Scalettar, *Pairing and spin gap in the normal state of short coherence length superconductors*, Phys. Rev. Lett. **69**, 2001 (1992).

83  N. Trivedi, R. T. Scalettar, and M. Randeria, *Superconductor-insulator transition in a disordered electronic system*, Phys. Rev. B **54**, 3756 (1996).

84  E. W. Huang, R. Sheppard, B. Moritz, and T. P. Devereaux, *Strange metallicity in the doped Hubbard model*, arXiv:1806.08346 (2018).

85  S. Lederer, Y. Schattner, E. Berg, and S. A. Kivelson, *Superconductivity and non-fermi liquid behavior near a nematic quantum critical point*, Proc. Natl. Acad. Sci. (U.S.A.) **114**, 4905 (2017).

### 7. Atomically precisely doped semiconductors

Jonathan Wyrick[1] and Shashank Misra[2]

[1]*National Institute of Standards and Technology,* [2]*Sandia National Laboratories*

The discovery of scanned probe microscopy thirty-five years ago opened the door to real-space imaging of atoms for the first time.[1] After early successes in fabricating company logos with atomic precision, only recently have workflows produced useful electrical devices, using scanned probe fabrication methods.[2] In particular, the use of hydrogen lithography[3,4] to create donor-based devices in silicon[5] has seen a surge in activity related to demonstrating the limits of scaling circuit elements to the physical limit of atoms themselves,[6,7] and for demonstrating control over the charge and spin degrees of freedom of islands containing a small number of donors.[8,9] This workshop explored the possibility that these dopant islands can serve as a kind of artificial atom, and thus finite-size arrays should exhibit Bloch bands emerging inside the bandgap of silicon from the overlap of shallow dopant levels; and, in certain limits, strong electronic correlations.

The promise and limitations of the resultant quantum system can be understood by considering how the artificial atoms themselves are fabricated, and from their resultant properties.

Fabrication proceeds by taking the (100) surface of silicon and cleaning it in ultra-high vacuum, which creates a $2 \times 1$ dimer surface reconstruction.[10] With this surface reconstruction, each surface silicon atom is left with one dangling bond. Attachment of a hydrogen atom to these dangling bonds renders the surface unreactive. The scanning tunneling microscope (STM) can, at moderate junction conditions, be used to inelastically break the silicon – hydrogen bond.[11] At low junction bias, STM stops removing hydrogen, and can image both structural elements of the surface, and electronic features like passivated versus unpassivated sites. (see Fig. 11) Hydrogen lithography produces a surface with selective chemical reactivity, which can be done with single-site precision using STM.

Phosphine selectively adsorbs only on unterminated sites and is self-limited to a monolayer. A thermally-activated decomposition of the phosphine at moderate temperatures produces an electrically active substitutional phosphorus dopant.[12] Phosphine adsorbs into an open dimer by decomposing into $PH_2$ + H. It is believed that three consecutive dimers must adsorb phosphine molecules for the subsequent decomposition reaction to incorporate a single phosphorus, which produces a positional uncertainty of just $\pm$ 1 lattice constant ($\sim$ 0.38 nm). Fortunately, this chemical pathway error corrects imperfections of the hydrogen resist, which are manifest mostly as single dimers missing terminations, and are not observed to cluster near room temperature. Windows smaller than 3 dimers thus do not incorporate any phosphorus.[13] The phosphorus dopants, called donors because they donate an electron into silicon, behave as artificial atoms with renormalized parameters stemming from the host silicon lattice.

The properties of these artificial atoms make this platform an attractive one for analog quantum simulation of difficult condensed matter problems.[14] This is most easily seen by treating arrays of these artificial atoms like a 2D Fermi–Hubbard lattice. (see Fig. 12) The cost in energy of removing or adding an electron to a neutral donor, the charging energy, has been directly measured to be about 50 meV based on experiments on a single-atom transistor device.[15] This corresponds to the Hubbard $U$ parameter. A second critical parameter relates to the tunnel coupling between two donors. This too has been measured

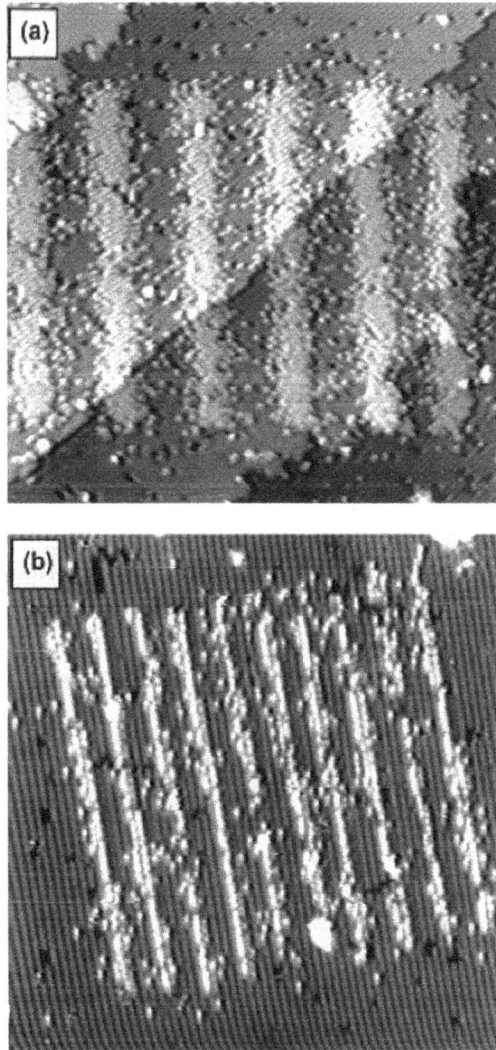

Fig. 11. STM images of hydrogen lithography. (a) Field emission depassivation at a junction of 7 V, 0.5 nA. (b) Nearly single-dimer lines written at 4.5 V, 2 nA. Diagonal line in (a) is an atomic step. From Ref. [4]

using STM spectroscopic imaging of two adjacent dopants and was found to be of around 5 meV at a donor separation of 3 nm.[16] This corresponds to the Hubbard $t$ parameter. Placing these samples in a dilution refrigerator, with base temperature of $\sim 100$ mK (10 μeV), puts donors in the regime where $t/T \sim 500$, $U/T \sim 5000$, and $U/t \sim 10$. (Ref. [17])

Designed arrays of dopant-based islands could establish an engineered approach to Hubbard model physics. Fabricating arrays with different periods that systematically vary $t$, while fabricating arrays with different island sizes systematically varies $U$. At the same time, the symmetry of the array can be varied — e.g. from square to triangular — with simple changes in lithographic patterns.

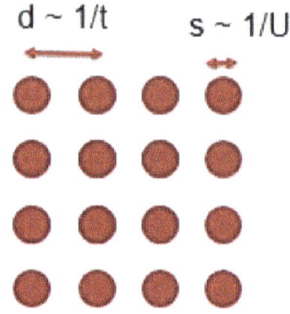

Fig. 12. Schematic of a 4 × 4 array of donor islands. Hubbard $U$ decreases with larger island size $s$. Hubbard $t$ decreases with larger array periodicity $d$.

This stands in contrast to the traditional mode of exploration of quantum materials, where different compounds must be synthesized and purified to change the Hubbard parameters, often in an Edisonian fashion. Such an array is meant to serve as a direct one-to-one analog of difficult problems in quantum materials. An array of dopants with the aforementioned parameters is in the strongly correlated, low temperature limit, where we expect new quantum phases to emerge. (see Fig. 13) In a square Hubbard array, we expect the strongly — correlated Mott phase to emerge when $T < t^2/U \sim 10$ K. Moreover, numerical modeling of an 8 × 8 Hubbard array shows robust superconductivity at $T_c \sim t/40 \sim 1$ K. (Ref. [18])

In comparison to cold atom experiments, donors have a well-defined and uniform temperature, larger energy scales relative to system temperature, and a natural representation with electrons. However, cold atoms are identical and the arrays, being defined by laser fields, are nearly perfect. By contrast, dopants and semiconductors are nowhere as clean (i.e. identical). Fabricated systems always have more disorder than nature. For phosphorus donors, the

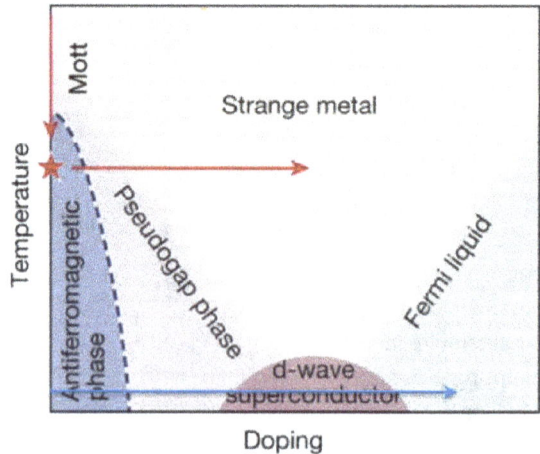

Fig. 13. The high-$T_c$ superconductivity phase diagram, which is believed to correspond to a large n Hubbard model. The red line is the operating point of the trapped atom experiment in Ref. [17]. The blue line is the expected operating region for a donor array in a dilution refrigerator, for which $t/T \sim 500$, $U/T \sim 5000$, and $U/t \sim 10$

stochastic nature of the phosphine incorporation chemistry leads to disorder. A lithographic window of 3 × 2 dangling bonds only incorporates a phosphorus roughly 70% of the time (see Fig. 14), with no dopant incorporation 30% of the time. Based on these incorporation statistics, the probability that a lithographically perfect array of 5 by 5 single donor windows producing a dopant array with no missing elements is less than 0.01%. Even permitting some breakthrough where each node can be made to be a single donor, experience from color centers in semiconductors and insulators indicates that "identical" defects are never spectroscopically identical.

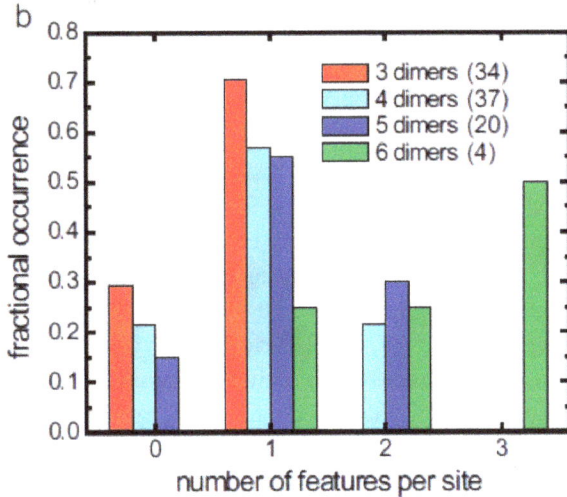

Fig. 14. Stochastic nature of dopant incorporation through phosphine decomposition in STM-depassivated lines of the indicated length. From [15]

A potential solution to this problem lies in defining many-donor islands. If each node of a 5 × 5 array consists of three adjacent 3 × 2 windows, the probability of producing an array with no missing elements is 50%, with the compromise that each node will have between 1 and 3 donors. The increased size of the islands will decrease Hubbard $U$, and will require a concomitant decrease in Hubbard $t$ (or increase in array spacing) to maintain the same ratio of energies. Roughly speaking, this is expected to modestly deteriorate characteristic energy scales by less than a factor of two, largely preserving the comparison of Fig. 13. More importantly, the resultant array consists of a set of nodes with a distribution of charging energies, and of tunnel couplings to neighboring nodes. A significant open question is the effect of this disorder, which could create filamentary conduction paths through the array. Calculations of disorder on 1D chains indicate correlations survive at low temperatures.[19] Another open challenge derives from the finite size of any array. The aforementioned 5 × 5 array has 16 perimeter (edge) and 9 interior (bulk) sites, and thus may not exhibit the expected bulk behavior of a Hubbard array. A second challenge arises from the expectation of an extended Coulomb potential for real donors. This can create a localization potential in the center of the array (sites with a larger number of nearest and next nearest neighbors).[20] On the other hand, quantum many body states (intuitively) may provide some inherent robustness against these kinds of disorder, which may doom their utility to more sensitive applications.

We expect that a given array, where Hubbard $U$ and $t$ will be fixed by the fabrication of $n \times n$ islands of size $s$ and pitch $d$, can be tuned between insulating and metallic ground states upon changing the occupancy away from half filling. The electronic state of the array can be simply queried by fabricating it between tunnel coupled leads and measuring transport through the array. Tunneling spectroscopy has been used in the past to interrogate the electronic structure of an isolated donor (Fig. 15), and, by extension, can be expected to probe something akin to the density of electronic states of the array. The integration of in-plane gates or surface gates can be used to change the electron occupancy of the array. Data from multiple samples, each with an array with different $s$, $d$, and $n$, can then be used to map crossovers between different states of finite sized Hubbard arrays and compared to calculated 'phase' diagrams (Fig. 16).

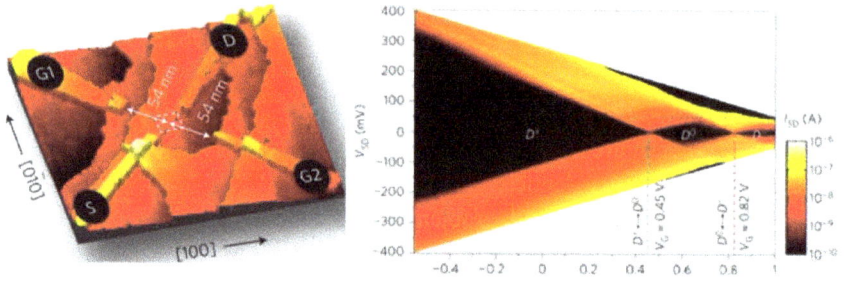

Fig. 15. (Left) STM image of hydrogen mask that produced a single phosphorus donor (middle of dashed box), which can be gated by G1 and G2 to yield resonant tunneling between source and drain. (Right) Electrical transport through the single atom transistor, showing clear spectroscopic identification of the charge states of the donor. From Ref. [7]

Overall, disorder effects, along diagonal hopping, and an extended Hubbard interaction, need to be calibrated to expectations based on comparing experimental results on smaller arrays to detailed numerical calculations. Work on smaller arrays is crucial to establish confidence in the behavior of larger arrays, where comparison to numerical modeling will require more approximations and assumptions.

If this combined approach validates an understanding that the dopant arrays provide a tunable system with emergent many-body physics, there are a range of opportunities where these arrays can be applied to useful physical problems that are close at hand. Changing the array symmetry immediately permits access to other physics, like magnetic frustration in a Mott state on a triangular lattice, or

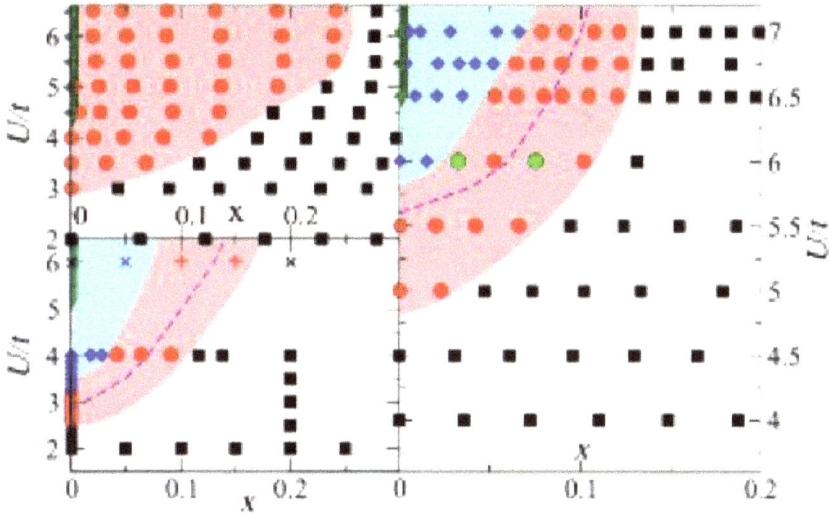

Fig. 16. Phase diagram of an 8 × 8 (right), 4 × 4 (upper left) and 16 × 16 (lower left) Hubbard array in terms of Hubbard parameters $U$, $t$, and lattice filling $x$. The blue region supports superconductivity. The green line near $x = 0$ denotes the correlated insulator, the red region denotes the superconductor, the blue region denotes a non-superconducting pseudogap phase, and the rest of the diagram contains a normal Fermi liquid. From Ref. [18]

strongly interacting Dirac electrons on a hexagonal lattice (Fig. 17). Across this broad range of lattice types, dopant arrays could be used to benchmark different approximations and numerical approaches against one another. A second opportunity derives from integrating measurements on larger arrays into theoretical frameworks based on quantum impurity models. Here, the donor arrays can be used as analog accelerators in numerical models of hard quantum material problems.[21]

Incorporation of precursor species beyond phosphorus expands the range of quantum materials where this technique applies. Patterning two interpenetrating arrays, each incorporating different atomic species, permits the analog quantum simulation of planar materials with a diatomic basis, like the dichalcogenides or Lieb materials (Fig. 17). Arrays where each node is a designed cluster can be used to introduce (and manipulate) orbital-like degrees of freedom. Precursors which incorporate magnetic species open the door for examining problems that map on to the Heisenberg Hamiltonian, instead of the Hubbard Hamiltonian.

Apart from applying arrays to wider classes of quantum materials problems, a wealth of opportunity lies in making measurements beyond resonant tunneling through the array to probe its electronic state (insulating or metallic). A simple extension of the measurement scheme is to measure the electrical response to different driving forces. A bowtie antenna could focus THz radiation to strongly

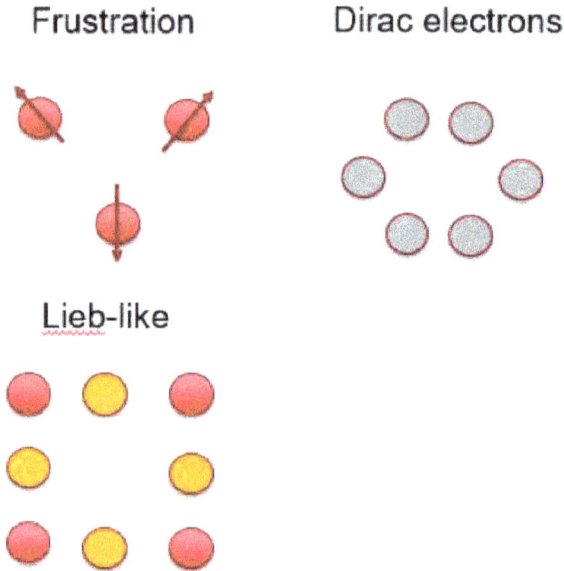

Fig. 17. (Top left) Mott state on a triangular array of donor islands, demonstrating magnetic frustration. (Top right) Donor arrays with hexagonal symmetry may be expected to contain Dirac-like quasiparticles. (Bottom left) Integration of a second dopant species may enable the creation of more involved lattices.

drive an array at energies as large as the tunnel coupling between adjacent sites. Perturbation of the array with a driving magnetic field may permit the measurement of a kind of magnetic susceptibility, important because strong correlations often result in interesting quantum many-body behavior in the spin channel instead of the charge channel. Further afield, measurement of other quantities, like the magnetization, seem necessary to realize the promise of the connection to condensed matter physics, but remain daunting when the system under test comprises 25 or 100 atoms.

**Acknowledgments**

Some work in this section was supported by the Laboratory Directed Research and Development program at Sandia National Laboratories, a multi-mission laboratory managed and operated by National Technology and Engineering Solutions of Sandia, LLC., a wholly owned subsidiary of Honeywell International, Inc., for the U.S. Department of Energy's National Nuclear Security Administration under contract DE-NA-0003525. This section describes objective technical results and analysis. Any subjective views or opinions that might be expressed in this section do not necessarily represent views of the U.S. Department of Energy or the United States Government.

## References

1    G. Binnig, H. Rohrer, C. Gerber, and E. Weibel, *7 x 7 Reconstruction on Si(111) resolved in real space*, Phys. Rev. Lett. **50** (2), 120–123 (1983). https://doi.org/10.1103/PhysRevLett.50.120.

2    C. Quate, *Scanning probes as a lithography tool for nanostructures*, Surf. Sci. **386**, 259–264 (1997). https://doi.org/10.1016/S0039-6028(97)00305-1.

3    J. W. Lyding, T.-C. Shen, J. S. Hubacek, J. R. Tucker, and G. C. Abeln, *Nanoscale patterning and oxidation of H-passivated Si(100)-2×1 surfaces with an ultrahigh vacuum scanning tunneling microscope*, Appl. Phys. Lett. **64**, 2010–2012 (1994). https://doi.org/10.1063/1.111722.

4    J. W. Lyding, K. Hess, G. C. Abeln, D. S. Thompson, J. S. Moore, M. C. Hersam, E. T. Foley, J. Lee, Z. Chen, S. T. Hwang, H. Choi, Ph. Avouris, and I. C. Kizilyalli, *Ultrahigh vacuum–scanning tunneling microscopy nanofabrication and hydrogen/deuterium desorption from silicon surfaces: Implications for complementary metal oxide semiconductor technology*, Appl. Surf. Sci. **130–132**, 221–230 (1998). https://doi.org/10.1016/S0169-4332(98)00054-3.

5    F. J. Ruess, L. Oberbeck, M. Y. Simmons, K. E. J. Goh, A. R. Hamilton, T. Hallam, S. R. Schofield, N. J. Curson, and R. G. Clark, *Toward atomic-scale device fabrication in silicon using scanning probe microscopy*, Nano Lett. **4** (10), 1969–1973 (2004). https://doi.org/10.1021/nl048808v.

6    B. Weber, S. Mahapatra, H. Ryu, S. Lee, A. Fuhrer, T. C. G. Reusch, D. L. Thompson, W. C. T. Lee, G. Klimeck, L. C. L. Hollenberg, and M. Y. Simmons, *Ohm's law survives to the atomic scale*, Science **335**, 64–67 (2012). https://doi.org/10.1126/science.1214319.

7    M. Fuechsle, J. A. Miwa, S. Mahapatra, H. Ryo. S. Lee, O. Warschkow, L. C. L. Hollenberg, G. Klimeck, and M. Y. Simmons, *A single-atom transistor*, Nat. Nano. **7** (4), 242–246 (2012). https://doi.org/10.1038/nnano.2012.21.

8    H. Buch, S. Mahapatra, R. Rahman, A. Morello, and M. Y. Simmons, *Spin readout and addressability of phosphorus-donor clusters in silicon*, Nat. Commun. **4**, 2017 (2013). https://doi.org/10.1038/ncomms3017.

9    T. F. Watson, B. Weber, Y.-L. Hsueh, L. C. L. Hollenberg, R. Rahman, and M. Y. Simmons, *Atomically engineered electron spin lifetimes of 30 s in silicon*, Sci. Adv. **3** (3), e1602811 (2017). https://doi.org/10.1126/sciadv.1602811.

10    R. M. Tromp, R. J. Hamers, and J. E. Demuth, *Si(001) dimer structure observed with scanning tunneling microscopy*, Phys. Rev. Lett. **55** (12), 1303–1306 (1985). https://doi.org/10.1103/PhysRevLett.55.1303.

11    T.-C. Shen, C. Wang, G. C. Abeln, J. R. Tucker, J. W. Lyding, Ph. Avouris, and R. E. Walkup, *Atomic-scale desorption through electronic and vibrational excitation mechanisms,* Science **268** (5217), 1590–1592 (1995). https://doi.org/10.1126/science.268.5217.1590.

12   R. Schofield, N. J. Curson, M. Y. Simmons, F. J. Ruess, T. Hallam, L. Oberbeck, and R. G. Clark, *Atomically precise placement of single dopants in Si*, Phys. Rev. Lett. **91** (13), 136104 (2003). https://doi.org/10.1103/PhysRevLett.91.136104.

13   O. Warschkow, N. Curson, S. Schofield, N. Marks, H. Wilson, M. Radny, P. Smith, T. Reusch, D. McKenzie, M. Simmons, *Reaction paths of phosphine dissociation on silicon (001)*, J. Chem. Phys. **144** (1), 014705 (2016). https://doi.org/10.1063/1.4939124.

14   I. M. Georgescu, S. Ashhab, and F. Nori, *Quantum simulation*, Rev. Mod. Phys. **86** (1), 153–185 (2014). https://doi.org/10.1103/RevModPhys.86.153.

15   M. Fuechsle, *Precision Few-Electron Silicon Quantum Dots*, Ph. D. Dissertation, School of Physics, University of New South Wales (2011).

16   J. Salfi, J. A. Mol, R. Rahman, G. Klimeck, M. Y. Simmons, L. C. L. Hollenberg, and S. Rogge, *Quantum simulation of the Hubbard model with dopant atoms in silicon*, Nat. Comm. **7**, 11342 (2016). https://doi.org/10.1038/ncomms11342.

17   A. Mazurenko, C. S. Chiu, G. Ji, M. F. Parsons, M. Kanasz-Nagy, R. Schmidt, F. Grusdt, E. Demler, D. Greif, and M. Greiner, *A cold-atom Fermi–Hubbard antiferromagnet*, Nature **545** (7655), 462–466 (2017). https://doi.org/10.1038/nature22362.

18   E. Gull, O. Parcollet, and A. J. Millis, *Superconductivity and the pseudogap in the two-dimensional Hubbard model*, Phys. Rev. Lett. **110** (21), 216405 (2013). https://doi.org/10.1103/PhysRevLett.110.216405.

19   A. Dusko, A. Delgado, A. Saraiva, and B. Koiller, *Adequacy of Si:P chains as Fermi–Hubbard simulators*, Npj Quant. Info. **4** (1), 1 (2018). https://doi.org/10.1038/s41534-017-0051-1.

20   N. H. Le, A. J. Fisher, and E. Ginossar, *Extended Hubbard model for mesoscopic transport in donor arrays in silicon*, Phys. Rev. B **96** (24), 245406 (2017). https://doi.org/10.1103/PhysRevB.96.245406.

21   B. Bauer, D. Wecker, A. J. Millis, M. B. Hastings and M. Troyer, *Hybrid quantum-classical approach to correlated materials*, Phys. Rev. X **6** (3), 031045 (2016). https://doi.org/10.1103/PhysRevX.6.031045.

# 8. Cold atoms and optical lattices

Bhuvanesh Sundar and Kaden Hazzard

*Rice University*

## Overview of ultracold matter

It is useful to view ultracold matter in the context of atomic, molecular, and optical (AMO) physics' long tradition of manipulating quantum matter with increasing levels of control, illustrated in Fig. 18. In the 1800s and early 1900s, experiments were measuring the quantized electronic spectra of gases of atoms, as shown in Fig. 18(a). These measurements ushered in quantum mechanics, as just one of their profound impacts. The ability to precisely measure spectra naturally lead to the ability to control atoms' electronic states, pumping them into desired excited levels. The invention of the laser in the late 1950s and 1960s revolutionized the ability of physicists to control these excitations, allowing desired electronic states

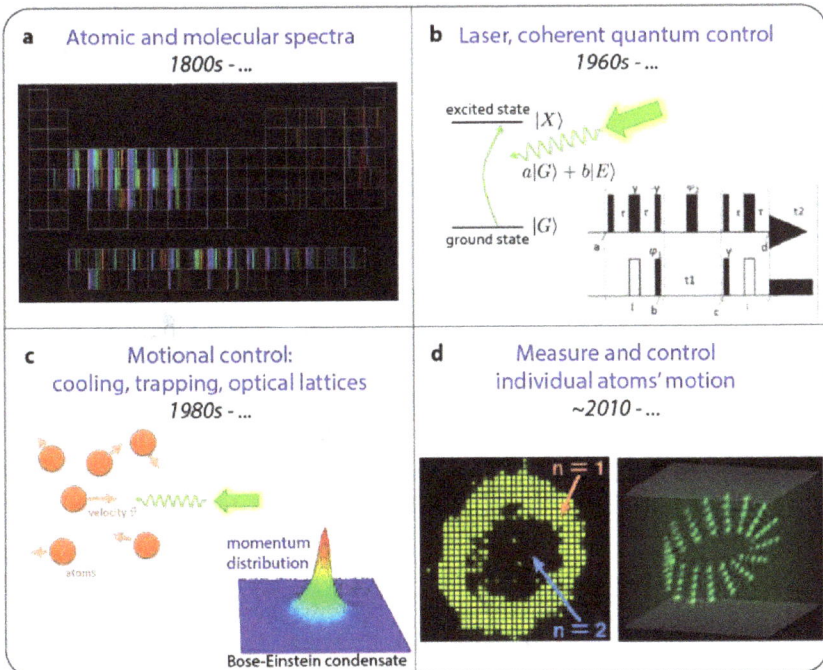

Fig. 18. Increasing control of quantum matter in AMO physics through history. (a) Emission spectra of the elements. (From Ref. [1]) (b) Control of coherent quantum superpositions of electronic levels. (Adapted from Ref. [2]) (c) Controlling motion of many atoms: laser and evaporative cooling, optical trapping, and new states of matter such as Bose-Einstein condensates. (Momentum distribution adapted from Ref. [3]) (d) Measuring and controlling many-body motional states atom-by-atom: quantum gas microscope (left, adapted from Ref. [4]) and programmable optical tweezer arrays (right, adapted from Ref. [5]).

to be populated, including high-fidelity quantum mechanical superpositions of different energy levels [Fig. 18(b)].

Lasers can control not only the internal state of the atoms, but also their motion in space, as demonstrated[b] in the 1980s.[6-8] This new layer of control, specifically the techniques to cool and trap atoms that are now universally used in ultracold experiments (Fig. 18(c)), ushered in a new era. A landmark application of these tools was to combine them with evaporative cooling[c] to create a new state of matter, dilute Bose–Einstein condensates (BECs), in 1995.[9] In this state, despite being dilute — a million times less dense than air — the indistinguishable atoms overlap and quantum statistics play a major role. Bosonic statistics compel the particles to coalesce into identical quantum states, at roughly the lowest single-particle energy state. This was the first example of many-body quantum degenerate matter in ultracold systems.

A great virtue of studying matter at these low temperatures and densities is that the atoms move so slowly that they are susceptible to the tiny forces generated by light and magnetic fields. Such forces are used to generate traps for the atoms as well as the lattices that are crucial to realize the Fermi–Hubbard model. These forces are now used to image and control many-body quantum systems atom-by-atom.[10,11] The exquisite level of control of these systems makes them natural candidates for *quantum simulators* of strongly correlated matter.

Henceforth, we will concentrate on quantum simulations of the Fermi–Hubbard model with ultracold fermions in optical lattices. Our reasons for focusing on this case are not that it is the only example of a model simulated in ultracold matter, or even the most important or exciting. In actuality, numerous states of strongly interacting matter and phenomena have been explored in ultracold matter. Several are mentioned in Sec. I (2.2), and even these encompass only a tiny fraction of those studied in experiments.

Rather, there are two important reasons we focus on the Fermi–Hubbard model. First, it is the most relevant point of comparison for the 2D Quantum Metamaterials Workshop and it has some of the most familiar connections to strongly correlated condensed matter. Second, exploring this model and its ultracold realizations will illuminate many general themes of ultracold matter — its successes, limitations, and potential.

A comprehensive and up to date review focused on quantum simulation of the Fermi–Hubbard model in ultracold matter can be found in Ref. [12].

---

[b] Cooling and trapping of neutral atoms were demonstrated in the 1980s. In trapped ions, laser cooling had been proposed and implemented in 1978.[7,8]

[c] Fundamentally, this is the same process that cools a cup of hot coffee to room temperature. Hot atoms evaporate out of the trap, which has a finite depth. In practice, experiments tune the depth of the trap as a function of time to optimize this evaporation efficiency.

The rest of this addendum material is organized as follows. The first section provides background on what ultracold matter is, typical length and energy scales, and important measurement techniques. The next section discusses how the Fermi–Hubbard model arises in these systems. Next, the section "Achievements of cold atom Fermi–Hubbard experiments" outlines the main achievements of ultracold realizations of the Fermi–Hubbard model in the context of quantum simulation. The final section discusses next steps, challenges, and prospects for the future.

**A primer on ultracold matter**

Ultracold matter refers to dilute gases of atoms or molecules trapped in high vacuum and at very cold temperatures. Broadly, the goal of most experiments on ultracold matter is to simulate phenomena with a high degree of control. To avoid unwanted interactions of atoms/molecules with the environment and with each other (such as chemical reactions), experimentalists work with a dilute gas that is isolated from the environment. A typical experiment has about $10^5$ particles in a small cloud roughly 100 μm in size in the middle of a vacuum chamber, resulting in a density of $\sim 10^{12}$ cm$^{-3}$. Because the density is so low (for comparison, 7 orders of magnitude lower than air), the temperature of the gas should also be low to observe quantum effects. The temperature for this is such that the inter-particle spacing is around or smaller than the thermal de Broglie wavelength of the atom/molecule, $n^{-\frac{1}{3}} \sim \lambda_{th}$ , yielding a nano-Kelvin scale temperature. Cooling gases to such a low temperature is the price that experimentalists pay to achieve precise control and eliminate unwanted interactions.

The popular workhorses for ultracold matter are the alkali atoms — lithium, sodium, potassium, rubidium, and cesium, owing to their simple electronic structure; all of them have one valence electron. Many of the alkali have multiple isotopes some of which are bosonic and some fermionic. Working with different atomic species requires different experimental conditions. However, researchers have cooled all these atomic species to nano-Kelvin scale temperatures (see Refs. [13-16] for reviews). New atomic species are being added to the list at a rapid pace, which now includes alkaline-earth-like atoms such as strontium and ytterbium,[17,18] as well as open-shell lanthanides such as erbium and dysprosium.[19,20] The development of techniques to create degenerate quantum gases of all these atomic species is driven by each one's unique advantages — some have simple electronic structures, while some others may have highly metastable excited states, and yet others may have strong magnetic dipole interactions. Experimentalists often work with atoms in the lowest hyperfine state, although higher hyperfine states or even excited electronic states may be used as well. The Zeeman states in the lowest hyperfine manifold act as a pseudospin, analogous to the spin of an electron in condensed matter physics. Other forms of ultracold matter of recent interest are alkali atoms in highly excited Rydberg states ($n > 50$), and ground-state bi-alkali molecules such as KRb, NaRb, NaK, and RbCs.[21-25]

In a typical experiment, a gas of atoms is emitted from a source such as a chunk of alkali metal in an oven, cooled in several stages such as laser cooling and evaporative cooling, and then trapped in vacuum by magnetic fields and/or light. An important question in all experiments on ultracold matter is the time scale in the experiment. In principle, a dilute gas of ultracold atoms is never in equilibrium, as the ground state is always a chunk of metal. Nevertheless, the gas is metastable over the duration of the experiment, which may be a few seconds, and therefore can attain thermodynamic equilibrium within the metastable phase space.

After cooling and trapping a gas, researchers couple the particles to external fields that are appropriately tailored to study the desired physics. The most common ways to interact with atoms/molecules is electromagnetic radiation (UV, visible, microwave, and rf) and static magnetic fields (up to several hundred Gauss). For example, researchers may load the particles into a lattice created by a standing wave of light to realize Hubbard or other lattice models. Researchers may also perform several experiments that do not involve a lattice, but instead require a homogeneous ultracold gas. Magnetic fields may be used in an experiment to trap atoms, as well as tune the interaction strength between atoms via a Feshbach resonance.[26]

An experiment typically concludes by taking an image of the gas, which measures the density of the gas at different points in space. Here, researchers use probing techniques that are unique to ultracold atoms. In one technique called "time-of-fight measurement", researchers turn off the trap and let the gas expand for a few hundred ms, and then capture an image using a camera. The time-of-flight measurement yields information about occupation in different momentum states; and, for example, straightforwardly discerns condensate and insulating phases. Another measurement technique that was developed recently for experiments on lattices is quantum gas microscopy,[10] which measures the number of atoms on each lattice site in-situ. Developments in quantum gas microscopy have enabled researchers to study various phenomena, such as phase diagram of the Bose–Hubbard model,[27] localization in the Aubrey–Andre model,[28-30] and a direct measurement of the entanglement entropy,[31] as just a few examples, as well as the physics of the Fermi–Hubbard model, as we'll see in more detail below.

**How the Fermi-Hubbard model is realized in ultracold matter**

In this section, we describe how the Fermi–Hubbard model arises as a quantitative description of ultracold fermions in optical lattices,[32] and how experiments achieve their high degree of control of the lattice and system parameters. Experiments can make artificial crystals — specifically, periodic potentials for the atoms — known as optical lattices by interfering laser beams,[9] which control the tunneling and on-site interaction between atoms by tuning the laser intensity. Additionally, the interactions can be controlled by tuning a magnetic field. This allows experimentalists to accurately simulate the Hubbard model over a wide range of tunneling-to-interaction ratios.

In contrast to naturally occurring solid-state systems, the Hubbard model is not simply a toy model that captures some essential physics. Rather, it is a quantitatively accurate, rigorously derivable model for atoms in optical lattices in the ultracold regime. Here, we explain in detail the ingredients to simulate the Hubbard model for cold atoms, and from these ingredients derive the Hubbard model as a description of the system.

To create a potential for atoms using light, experimentalists use the AC Stark shift[9] that results from coupling to light that is far detuned from an atomic transition. To understand this, consider an electromagnetic field of frequency illuminating a two-level atom with an energy difference $\hbar\omega_0$ between the two levels, as shown in Fig. 19(a). The detuning of the laser from the excited state is $\delta = \omega_0 - \omega$. The light does not couple to any other atomic levels that are nearby in energy, and therefore to a good approximation, the atom can be assumed to be a two-level system. The atom may also carry a pseudospin index, which we will introduce later. The Hamiltonian for the system is

$$\hat{H} = \hbar\omega_0 \, |e\rangle\langle e| + \hat{H}_{atom-light},$$

where $\hat{H}_{atom-light}$ describes the coupling between the atom and light, given by

$$\hat{H}_{atom-light} = \hat{\vec{d}} \cdot \vec{E} \cos(\omega t) = d_{eg}E \cos(\omega t) \, (\,|g\rangle\langle e| + |e\rangle\langle g|),$$

with $\hat{\vec{d}}$ the dipole operator and $\vec{E}$ the electric field amplitude.

In the frame rotating at a frequency $\omega$, the rotating wave approximation is highly accurate and yields the effective Hamiltonian

$$\hat{H} = \Omega(|e\rangle\langle g| + |g\rangle\langle e|) + \hbar\,\delta\,|e\rangle\langle e|,$$

where $\Omega = \frac{d_{eg}E}{2\hbar}$. Finally, in the limit where the detuning $\delta \gg \Omega$, we eliminate the excited state $|e\rangle$ using second-order perturbation theory, and obtain the low-energy Hamiltonian

$$\hat{H}_{eff} = -\frac{\hbar\,\Omega^2}{\delta} \, |g\rangle\langle g|.$$

This is called the AC Stark shift.[9] It is a potential energy for an atom that is proportional to the intensity of the laser, and inversely proportional to the detuning of the laser from atomic transition from $|g\rangle$ to $|e\rangle$.

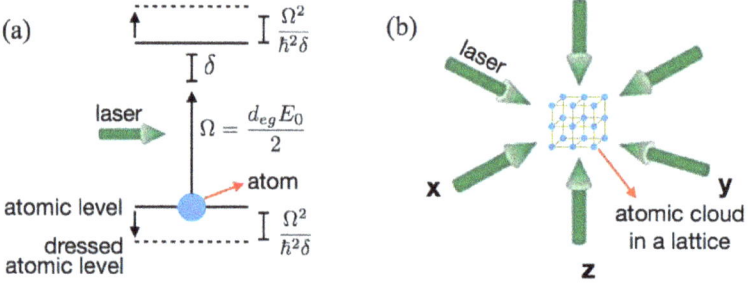

Fig. 19. (a) Illustration of the AC Stark shift for a two-level atom illuminated by far-detuned light. The energy of the dressed atomic levels is shifted from the bare atomic levels by an amount proportional to the intensity of the light. (b) Schematic of a laser configuration used to create a cubic optical lattice. Other lattices may be created by adjusting the laser configuration.

Experimentalists create an optical lattice for the atoms by interfering two counter-propagating laser beams.[9] The intensity of the interference pattern modulates periodically in space, resulting in a periodic potential for the atoms. The effective Hamiltonian for the atoms (not yet including the interaction) is

$$\hat{H} = \sum_\sigma \int d^3\vec{r}\, \hat{\psi}_\sigma^{\,\dagger}(\vec{r}) \left( -\frac{\hbar^2 \nabla^2}{2m} + V(\vec{r}) \right) \hat{\psi}_\sigma(\vec{r}) \,,$$

where $V(\vec{r}) = V_0 \cos^2(\vec{k} \cdot \vec{r})$ is the periodic potential created by the interfered laser beams, $V_0$ is the lattice depth proportional to the intensity of the lasers, $\vec{k}$ is the wave vector of the periodic modulation, and $\hat{\psi}_\sigma(\vec{r})$ is the annihilation operator for an atom with spin $\sigma$ at $\vec{r}$. A schematic of the setup is shown in Fig. 19(b).

The periodic potential created in this manner is free of defects or dislocations. Different lattices can be created by tuning the laser configurations, and their directions and polarizations; e.g. researchers can create a 1D lattice, 2D lattices such as square, triangular,[33,34] honeycomb,[34,35] Kagome,[36] 3D lattices such as a cubic lattice and various kinds of superlattices.[37-40]

The Hamiltonian that describes atomic motion in a deep optical lattice (treating only the non-interacting part) is the tight-binding Hamiltonian. To see this, write the field operator as $\hat{\psi}_\sigma(\vec{r}) = \sum_i \phi_i(\vec{r}) \hat{a}_{i\sigma}$, where $\phi_i(\vec{r})$ is the Wannier function for the lowest Bloch band localized at the lattice site labeled $i$, and $\hat{a}_{i\sigma}$ annihilates an atom of spin $\sigma$ from the lowest Bloch band at the site $i$. Substituting the field operator in the Hamiltonian, we obtain[32]

$$\hat{H}_{TBM} = -t \sum_{\sigma,<ij>} \hat{a}_{i\sigma}^{\dagger} \hat{a}_{j\sigma} \,,$$

where

$$t = -\int d^3\vec{r}\, \phi_i^*(\vec{r}) \left( -\frac{\hbar^2 \nabla^2}{2m} + V(\vec{r}) \right) \phi_j(\vec{r}) \,.$$

Here, $t$ is the tunneling amplitude, and the sum in $\hat{H}$ runs over neighboring sites $i$ and $j$. The tunneling amplitude can be directly controlled by the laser intensity and/or detuning. For example, increasing the lattice depth localizes the Wannier functions more, thereby decreasing $t$. In principle, longer-range tunneling and operators involving higher bands are present. However, the next-nearest-neighbor tunneling amplitude is typically small due to negligible overlap between those Wannier functions. And because the temperature (and interactions treated below) are much smaller than the band spacing, occupation of higher bands is negligible.

In addition to the tight-binding Hamiltonian above, the atoms have van der Waals interactions. The range of this van der Waals interaction, $\sim$ few nm, is much smaller than typical inter-atomic spacing, $\sim\mu$m. In this limit, the van der Waals interaction is well-approximated by a delta function pseudopotential, $V_{int}(\vec{r}) = (4\pi\hbar^2 a_s/m)\delta(\vec{r})$, characterized by a single parameter — the scattering length $a_s$. Similar to the non-interacting Hamiltonian, we can project the interactions on the lowest-band of the lattice, and only retain the on-site terms. For a pseudospin-½, this leads to an effective on-site interaction between the atoms,[32]

$$\hat{H}_{int} = U \sum_i n_{i,\uparrow} n_{i,\downarrow} \, ,$$

where $n_{i,\sigma} = a_{i,\sigma}^\dagger a_{i,\sigma}$ , and

$$U = \frac{4\pi\hbar^2 a_s}{m} \int d^3\vec{r} \, |\phi_i(\vec{r})|^4 \, .$$

The interaction strength $U$ can be controlled by tuning the lattice depth. For example, increasing the lattice depth localizes the Wannier functions more and increases the overlap of the Wannier function on a site with itself, thereby increasing $U$. Experimentalists can controllably tune $U/t$ over several orders of magnitude by adjusting the lattice depth. Another technique to tune $U$ is to adjust the scattering length $a_s$, achieved by tuning a magnetic field near a Feshbach resonance, which shifts the energy of a two-atom bound state relative to the energy for unbound atoms.

Together with $\hat{H}_{TBM}$ and $\hat{H}_{int}$, the total Hamiltonian for the atoms in an optical lattice is the Hubbard model,

$$\hat{H}_{Hubbard} = \hat{H}_{TBM} + \hat{H}_{int} \, .$$

## Extensions to the Hubbard model

In addition to producing the Hubbard model given above, experiments on ultracold atoms also enjoy the luxury of adding more ingredients to the system in a controllable manner. Some pursuits that have become popular recently have been the addition of strong artificial magnetic fields, disorder with controllable strength, higher Bloch bands in the lattice, SU(N)-symmetric interactions, and long-range interactions, to name a few. Here, we elaborate on some of the ideas to add these ingredients.

Neutral atoms do not feel a Lorentz force from a real magnetic field, an essential ingredient to simulate phenomena such as the Harper–Hofstadter model,[41] and integer and fractional quantum Hall effects.[42] Therefore, researchers devise schemes to produce strong artificial magnetic fluxes for cold atoms by coupling them to light that has been tailored accurately. A large magnetic flux per plaquette is achieved[43-45] by inducing complex hopping amplitudes between lattice sites, which introduces a Peierls phase for the atoms. To accomplish this, researchers turn on a lattice tilt along one direction, say the $\hat{x}$ direction, so that hopping is suppressed along $\hat{x}$. Then, researchers reintroduce hopping along $\hat{x}$ by shining two Raman lasers that drive photon-assisted tunneling; an atom absorbs and emits a photon to tunnel from one site to the next. In the process of absorption and emission, the atom also picks up a complex phase due to the spatial phase of the electromagnetic field. The magnetic flux per unit cell induced in this manner can reach values as large as the flux quantum $h/2e$, which would require applying more than 1000 Tesla in a solid-state system.

Researchers can introduce quasi-randomness in optical lattices by superposing two optical lattices with incommensurate periodicities.[28,30] This technique has been used to study many-body localization of fermions in 1D systems, as well as opening prospects to study many-body localization in higher dimensions. In fact, researchers have already been exploring this phenomenon in two dimensions for bosons.[29]

Atoms can be controllably excited or coupled to the higher Bloch bands of an optical lattice by various means, such as shaking the lattice,[46,47] resonant transfer by introducing an energy bias between sites,[48] and two-photon Raman transitions.[49] These schemes allow researchers to study unconventional forms of superfluidity.

It is possible to go beyond the Hubbard model interactions by performing experiments with different kinds of ultracold matter. For example, in alkaline earth atoms, which have a large nuclear spin, the on-site interaction between atoms have full SU($N$) symmetry,[17] where $N$ can be as large as 10 for some species such as $^{87}$Sr. Using these species, it is possible to simulate the SU($N$) Heisenberg model, and other SU($N$)-symmetric Hamiltonians, which have exotic ground states such as spin liquids and valence bond solids.[17] Other kinds of many-body ultracold matter in optical lattices, such as ultracold molecules,[21-25] Rydberg atoms[50-55] and dipolar atoms such as chromium or some lanthanides[56-58] have long-range dipole interactions extending beyond the on-site Hubbard-type interactions. These long-range dipole interactions potentially have many applications as well, producing new phases such as density wave orders and $p$-wave superfluids, to multi-qubit operations in quantum computation.

## Achievements of cold atom Fermi–Hubbard experiments

To go from experimentally achieving a BEC to realizing the Fermi–Hubbard model, experimentalists needed two crucial ingredients: an optical lattice using the techniques discussed in the previous section and ultracold fermions. To explore the exciting physics of a lattice system near the Mott insulator phase, it is crucial to work in a regime that is cold compared to the band spacing and at a filling of roughly one atom per site. Each of these ingredients represent substantial lines of research, with broad impacts in their own right.

Researchers first created bosons at temperatures and densities where the Mott insulator physics manifests in 2002.[59] By varying the lattice depth to change the ratio of interaction to tunneling, they observed the system transition from a bosonic Mott insulator to a superfluid. The Mott insulator occurs when interactions are strong relative to the tunneling and the system is near an integer number of particles per site. Then the energy cost for an atom to tunnel, causing a deviation of the local density from its integer value, will suppress the motion of the atoms, leading to an insulating state. These conditions paved the way for experiments probing numerous properties (momentum distributions, real space correlations with single site resolution, spectra, transport properties, …) across the quantum phase transition under various conditions. The variety of situations has included 1D, 2D, and 3D lattices, as well as controllable disorder that can be turned on or off at will. These experiments thus proved to be, as suggested in Ref. [60], powerful quantum simulators of the Bose–Hubbard model, a "toy" model that had been introduced for granular superconductors more than a decade prior.[61]

Contemporaneously, experimentalists produced the first quantum degenerate gases of fermionic atoms in 1999. These atoms were in a mixture of two hyperfine states, thus forming effective spin-½ systems. It was necessary to use multiple spin states to enable the collision processes that are essential for equilibration during the cooling process, since identical fermions lack $s$-wave collisions and thus are effectively non-interacting at the low temperatures being studied. Fortuitously, spin-½ fermions are also the essential building blocks of many interesting strongly-correlated phenomena, such as superconductivity and the Fermi–Hubbard model.

Besides the Fermi–Hubbard experiments that we will focus on, these degenerate Fermi gases led to exciting new fields in their own right. For example, attractive Fermi gases with interactions controlled by a Feshbach resonance were used to create analogs of superconductors in 2004–2005, as reviewed in Ref. [62]. They subsequently have been used to study strongly correlated superconductivity, for example the BEC-BCS crossover and exotic superconductors, such as the Fulde–Ferrell–Larkin–Ovchinokov (FFLO) state.

Realizing the Fermi–Hubbard model by combining ultracold fermions with optical lattices was a natural goal for the community, with potential for huge scientific payoffs. This was driven in large part by the scientific interest in the Fermi–Hubbard model, which stemmed from two sources. First is the Fermi–Hubbard model's connections to high-temperature superconductivity, with it forming a minimal model often believed to capture essential physics of the cuprate superconductors. The second is the Fermi–Hubbard model's status as a standard model for strongly correlated physics that lies at least partially out of the reach of well-controlled numerical methods.

The first efforts loaded fermionic atoms into optical lattices and observed the interplay of quantum statistics with the band structure. Specifically, researchers[63] observed that spinless fermions with one particle per site form a band insulator, and (at lower densities) a metal, as shown in Fig. 20(a). These experiments were an exciting new direction for ultracold matter, but much of the essential physics did not involve interactions, and they were far from the Mott regime.

The first experiments in ultracold matter to explore Mott physics in the Fermi–Hubbard model were in 2008, in the groups of Tilman Esslinger[64] and Immanuel Bloch.[65] These experiments revealed the finite-temperature crossover from a metal to a Mott insulator. Two of the primary observations were a suppression of double occupancies and a significant dip in the compressibility when the system was near half-filling (one atom per site) and $U/t$ was large. More recent measurements of this are shown in Fig. 20(b). Similar to the Bose–Hubbard model, in this limit, there is a gap to density excitations from the state with one atom per site, and this gap was also observed in spectroscopic measurements by modulating the lattice depth. An important feature in all of these experiments is the presence of the trap, which means the density varies throughout the system, starting largest in the center and dropping as one progresses to the edges. For a slowly varying trap, one can think of the system as being composed of a homogeneous subsystem with a large chemical potential in the center decreasing and becoming negative as one works out to the edge. The density will therefore decrease as one approaches the trap edge, but it will be nearly constant in the incompressible region of the trap.

Experiments on the Fermi–Hubbard model's Mott-metal crossover have proliferated. Experimentalists have measured accurately how the equation of state — i.e. the density as a function of chemical potential $n(\mu)$ — by measuring how the density varies in the trap, as well as double-occupancies and the nearest neighbor correlations.[64,65,66,67] In addition to these 3D cubic lattice experiments,

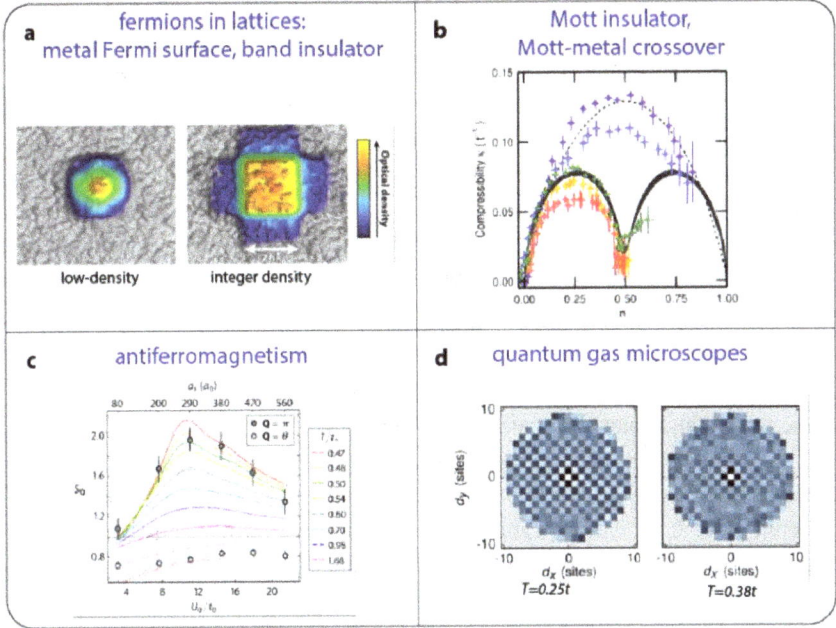

Fig. 20. Milestones in quantum simulations of the Fermi-Hubbard model with ultracold matter. (a) Observation of Fermi surfaces in lattices: momentum distributions measured by time-of-flight for two densities. (Adapted from Ref. [74]) (b) Precision measurements of Mott-metal equation of state for a number of interaction strengths from $U = 0$ to $U = 20\ t$ (different colored points) compared with non-interacting theory (dashed line) and $U = 8\ t$ theory from the dynamical cluster approximation (grey band), which agrees with the $U = 8\ t$ experimental data (green points). (Adapted from Ref. [69]) (c) Observation of strong antiferromagnetic correlations in 3D, above $T_N$. This shows the structure factor for the antiferromagnetic wavevector, $S(\pi)$, and zero wavevector, $S(0)$, as a function of $U/t$ compared with theory (solid lines connecting circles) at various temperatures. (Adapted from Ref. [81]) (d) Observations of real space spin-correlations $\langle S_i^z S_j^z \rangle - \langle S_i^z \rangle \langle S_j^z \rangle$ summed over all i and j, in a quantum gas microscope. (Adapted from Ref. [85])

similar experiments have been performed in 2D square lattices,[68-71] 2D honeycomb lattices,[72] and near 1D, as well as in anisotropic lattices interpolating between 1D and 3D.[73]

In addition to exploring this variety of lattice geometries, experiments have measured properties of generalizations of other regimes of Fermi–Hubbard models. They have explored the attractive ($U < 0$) Fermi–Hubbard model,[75] and the spin-imbalanced Fermi–Hubbard model.[76] Experiments with Yb atoms have created and studied Mott insulators in Fermi–Hubbard models with SU($N$) symmetric interactions for $N$ as large as $N=6$.[77,78] Experimentalists have also developed novel environments and probes, for example by coupling a 1D mesoscopic Fermi–Hubbard wire to fermionic reservoirs, making analogs of transport experiments possible.[79]

Although many experiments have studied the Mott insulator-metal crossover, the most interesting physics of the Fermi–Hubbard model also involves two other essential types of phenomena, with their own characteristic energy scales. (1) If the system is not exactly at half-filling, holes (sites without atoms) and doublons (sites with an extra atom) will tunnel, with characteristic energy $t$. (2) Spin correlations develop between particles in (or near) the Mott insulator. Specifically, being in a Mott insulator does not determine the behavior of the spin degree of freedom: as long as the system is dominated by configurations where each site harbors a single fermion, the spin degree of freedom of that fermion is free. However, there are effective spin-spin interactions. In the $U \gg t$ limit, these emerge from superexchange with a characteristic scale $t^2/U$. In the $U \ll t$ limit, these have a characteristic scale $U\, e^{-a\frac{U}{t}}$, for some constant $a$. In both cases, the effective interactions drive the system towards an antiferromagnetic state. In 3D systems, there is a finite temperature phase transition to an antiferromagnetically ordered Néel state at temperature $T_N$, while in lower dimensions there is finite (possibly very long) range order at finite temperature. In either case, the energy scales associated with these phenomena are often substantially less than $U$, and so, while achieving $k_B T \ll U$ suffices to experimentally explore the Mott-metal crossover, even lower temperatures are needed to explore physics associated with the spin correlations or the physics, such as potential superconductivity, arising when the density deviates from half filling.

In the last few years, experiments have begun to explore antiferromagnetic spin correlations associated with these smaller energy scales. This has been facilitated by a combination of colder temperatures and more refined experimental probes.[73,80-88] Two particularly important tools in this context are Bragg scattering and quantum gas microscopes.[10] The latter refers to experiments with the ability to measure every site of a 2D lattice, $i.e.$ the location of each atom in the lattice. This allows experiments to extract not only average densities with extreme spatial resolution, but also two-site (and higher) correlations of density and spin. Of particular importance is the observation of significant antiferromagnetic correlations, extending out to a range of several sites.

In particular, in 3D cubic lattices, experiments reached temperatures of roughly $T \sim 1.4\, T_N$ ($T \sim 0.5t$ and $U \sim 10t$) and measured the Bragg scattering signal at a wavevector commensurate with the expected antiferromagnetic ordering. They observed a substantial Bragg signal, which corresponded to a correlation length of roughly one lattice spacing [see Fig. 20(c)]. Put another way, a spin on one site is strongly correlated with its six neighbors and has non-negligible correlations with a larger shell.

In 2D square lattices, the strongest correlations and lowest temperatures have been reached in experiments that use quantum gas microscopes. Although true long-range order is forbidden at non-zero temperature for the SU(2) symmetry-breaking antiferromagnet in dimension two or lower, so $T_N = 0$, the longest-range antiferromagnetic correlations have in fact been observed in 2D. The longest ranged correlations have been reported in Ref. [85], who achieved temperature $T \leq 0.25\ t$. They observe non-negligible correlations across the whole system, roughly a circle containing 80 sites (*i.e.* a diameter of about 10 sites), with an exponential correlation length of about $\xi \sim 8$ sites [Fig. 20(d)].

It is worth noting that the majority of these experiments in both 2D and 3D are at or beyond the frontier of what can be quantitatively calculated numerically in a well-controlled fashion. For $T \sim t$ or higher, several numerical techniques can satisfactorily reproduce the equilibrium observables and equal-time correlations. Frequently used techniques include full exact diagonalization(ED),[89] determinantal quantum Monte Carlo (DQMC),[90] single-site dynamical mean-field theory (DMFT),[91] diagrammatic Monte Carlo,[92] and the numerical linked-cluster expansion (NLCE).[93] DQMC and NLCE have become the methods of choice, with DQMC often preferred for smaller $U/t$ and NLCE for larger $U/t$. However, at the frontier of temperatures $0.25t < T < t$, these numerical methods cease to be well-controlled, in the sense that there are numerical inaccuracies (e.g., finite-size effects) whose size may be significant, which are difficult to systematically evaluate under these conditions.

So far by judicious choice of algorithm and the employment of large-scale computer resources, numerical calculations have been able to reproduce experimental observations. However, due to the difficulty of directly assessing the numerical convergence explained above, the comparison with experiment provides arguably the best evidence that these methods are accurate in the regime of comparison. In this way, we are at the exciting time where experiments done in the present conditions give us information beyond the ability to be confidently extracted from numeric procedures. Note the discussion by Richard Scalettar and Ehsan Khatami in Sec. II (6).

**Prospects, challenges, and future directions**

*Future directions.* While ultracold experiments have already enhanced our understanding of the Fermi–Hubbard model, there are several clear next steps for these experiments with even greater impact. Achieving temperatures sufficient to reach the antiferromagnetically ordered state in 3D, *i.e.* $T < T_N$, is a major goal. Once achieved, this will allow experimentalists to apply the powerful tools of ultracold matter to studying a super-exchange-driven magnetically ordered phase, including the critical phenomena associated with the magnetic phase transition.

Moreover, once such temperatures are within reach, there are numerous parameters to vary to explore the different physics. For example, experiments could implement different lattice geometries to study frustrated magnetism, apply disorder and study its effects, and engineer topological band structures to investigate the interplay of topology and interactions. Diverse phenomena will become immediately accessible.

A long-time motivational goal for the field, and a logical next step after antiferromagnetic order is achieved, is to realize the $d$-wave superconducting phase that is widely predicted to occur in this model, or to rule out its existence. To do this, experiments must study the same system away from half-filling and at even lower temperatures than the Néel temperature, perhaps by a factor of 2 or 3. If such temperature can be reached, and suitable measurement tools applied, ultracold experiments will be able to directly probe a phase of this model that has been intensely debated for more than 30 years.

Perhaps the most interesting physics in the Fermi–Hubbard model may not even require achieving such low temperatures. Two phenomena of particular interest are possible pseudogap and the bad metal regimes of the Hubbard model. Studying these phenomena in ultracold matter may be possible in the near term, and in doing so would lead to profound new insights about quantum materials.

The first of these phenomena, the pseudogap, is displayed in many materials, which have a strongly suppressed density of states resembling an energy gap. There are many mechanisms that can give rise to a pseudogap, some mundane and some exotic.[94] The second of these phenomena, bad metals, occur in numerous materials, including high-temperature superconductors.[95] The question of what gives rise to the pseudogap and bad metal behavior in the cuprates (and other materials), specifically whether these behaviors are captured by the Hubbard model, is one that can potentially be addressed in ultracold experiments.

*Challenges.* Many of the phenomena described above require lower temperatures than have been demonstrated to date; but, encouragingly, the majority may occur within a factor of two or so of the current state of the art. The key challenge to reaching lower temperature is to remove entropy from the system faster than it is added by inevitable heating processes.

In present experiments, entropy is removed by evaporative cooling, which reduces the entropy by transporting high energy particles out of the trap. However, as experiments explore physics at lower energy and temperature scales, the characteristic timescales get longer, and entropy transport becomes very slow. This is compounded by the fact that it needs to be transported across the whole system, for example from the center to the edge. Given that the system may even be in an insulator, transport is essentially frozen over the lifetime of the experiment, as elegantly demonstrated in the Bose–Hubbard model.[96] Meanwhile, heating occurs due to inelastic collisions and light-scattering from lattice and trap lasers.[97] The key for progress is then finding regimes to minimize heating processes, and devising more efficient schemes to remove entropy than simple evaporative cooling. Speeding up transport, and in particular avoiding the necessity of redistributing entropy across long distances via slow transport processes, seems crucial.

Although there are challenges to overcome, they are not fundamental. With reasonable advances, we expect ultracold matter to reach into unprecedented regimes of the Fermi–Hubbard model and its relatives. In the process, we will learn even more about strongly correlated fermions. This will certainly include phenomena that are already of great interest — antiferromagnetism, $d$-wave superconductivity, pseudogap, and bad metal behavior. Moreover, there are dozens of models beyond the Fermi–Hubbard model and other phenomena being explored, each exciting in its own right. The biggest advances will surely be surprises that we have yet to anticipate.

### References

1   Rainbowkitteh, Emission Spectra of the Elements. took some time, [Online]. Available: http://rainbowkitteh.tumblr.com/post/83032827266/emission-spectra-of-the-elements-took-some-time.

2   H. Okamura, H. Nishimura, T. Nagata, T. Kigawa, T. Watanabe and M. Katahira, *Accurate and molecular-size-tolerant NMR quantitation of diverse components in solution*, Scientific Reports, **6**, 21742 (2016).

3   M. Greiner, C. A. Regal and D. S. Jin, Emergence of a molecular Bose–Einstein condensate from a Fermi gas, Nature, **426**, 537 (2003).

4   W. S. Bakr, P. M. Preiss, M. E. Tai, R. Ma, J. Simon and M. Greiner, *Orbital excitation blockade and algorithmic cooling in quantum gases*, Nature, **480**, 500 (2011).

5   D. Barredo, V. Lienhard, S. d. Léséleuc, T. Lahaye and A. Browaeys, *Synthetic three-dimensional atomic structures assembled atom by atom*, Nature, **561**, 79 (2018).

6   H. J. Metcalf and P. van der Straten, *Laser Cooling and Trapping*, New York: Springer-Verlager, (1999).

7   D. J. Wineland, R. E. Drullinger and F. L. Walls, *Radiation-Pressure Cooling of Bound Resonant Absorbers*, Physical Review Letters, **40**, 1639 (1978).

8   W. Neuhauser, M. Hohenstatt, P. Toschek and H. Dehmelt, *Optical-Sideband Cooling of Visible Atom Cloud Confined in Parabolic Well*, Physical Review Letters, **41**, 233 (1978).

9   C. J. Pethick and H. Smith, *Bose-Einstein Condensation in Dilute Gases*, Cambridge: Cambridge University Press, (2008).

10  S. Kuhr, *Quantum-gas microscopes: a new tool for cold-atom quantum simulators*, National Science Review, **3**, 170 (2016).

11  D. Barredo, V. Lienhard, S. d. Léséleuc, T. Lahaye and A. Browaeys, *Synthetic three-dimensional atomic structures assembled atom by atom*, Nature, **561**, 79 (2018).

12  L. Tarruell and L. Sanchez-Palencia, *Quantum simulation of the Hubbard model with ultracold fermions in optical lattices*, arXiv:1809.00571.

13  W. Ketterle, D. S. Durfee and D. M. Stamper-Kurn, *Making, probing, and understanding Bose–Einstein condensates*, in Proceedings of the International School on Physics Enrico Fermi 1998, *Bose–Einstein Condensation in Atomic Gases* , Amsterdam, IOS Press, **67**, (1999).

14  W. Ketterle and M. Zwierlein, *Making, porbing, and understanding ultracold Fermi gases*, in Proceedings of the International School on Physics Enrico Fermi (2006), *Ultracold Fermi Gases* , Amsterdam, IOS Press, **95**, (2007).

15  F. Dalfovo, S. Giorgini, L. P. Pitaevskii and S. Stringari, *Theory of Bose–Einstein condensation in trapped gases*, Reviews of Modern Physics, **71**, 463 (1999).

16  S. Giorgini, L. P. Pitaevskii and S. Stringari, *Theory of ultracold atomic Fermi gases*, Reviews of Modern Physics, **80**, 1215 (2008).

17  M. Cazalilla and A. Rey, *Ultracold Fermi gases with emergent SU( N ) symmetry*, Reports on Progress in Physics, **77**, 124401 (2014).

18  S. Stellmer, F. Schreck and T. C. Killian, Chapter 1: *Degenerate Quantum Gases Of Strontium*, in Annual Review of Cold Atoms and Molecules **2**, Singapore, World Scientific, **1**, (2014).

19  M. Lu, N. Q. Burdick, S. H. Youn and B. L. Lev, *Strongly Dipolar Bose–Einstein Condensate of Dysprosium*, Physical Review Letters, **107**, 190401 (2011).

20  K. Aikawa, A. Frisch, M. Mark, S. Baier, A. Rietzler, R. Grimm and F. Ferlaino, *Bose-Einstein Condensation of Erbium*, Physical Review Letters, **108**, 210401 (2012).

21  L. D. Carr, D. DeMille, R. V. Krems and J. Ye, *Cold and ultracold molecules: science, technology and applications*, New Journal of Physics, **11**, 055049 (2009).

22  B. Gadway and B. Yan, *Strongly interacting ultracold polar molecules*, Journal of Physics B: Atomic, Molecular and Optical Physics, **49**, 152002 (2016).

23  S. A. Moses, J. P. Covey, M. T. Miecnikowski, D. S. Jin and J. Ye, *New frontiers for quantum gases of polar molecules*, Nature Physics, **13**, 13 (2017).

24  M. Lemeshko, R. V. Krems, J. M. Doyle and S. Kais, *Manipulation of Molecules with Electromagnetic Fields*, Molecular Physics, **111**, 1648 (2013).

25  J. L. Bohn, A. M. Rey and J. Ye, *Cold molecules: Progress in quantum engineering of chemistry and quantum matter*, Science, **357**, 1002 (2017).

26  C. Chin, R. Grimm, P. Julienne and E. Tiesinga, *Feshbach resonances in ultracold gases,* Reviews of Modern Physics, **82**, 1225 (2010).

27  W. S. Bakr, A. Peng, M. E. Tai, R. Ma, J. Simon, J. I. Gillen, S. Fölling, L. Pollet and M. Greiner, *Probing the Superfluid-to-Mott Insulator Transition at the Single-Atom Level*, Science, **329**, 547 (2010).

28  M. Schreiber, S. S. Hodgman, P. Bordia, H. P. Lüschen, M. H. Fischer, R. Vosk, E. Altman, U. Schneider and I. Bloch, *Observation of many-body localization of interacting fermions in a quasi-random optical lattice*, Science, **349**, 842 (2015).

29  J.-Y. Choi, S. Hild, J. Zeiher, P. Schauß, A. Rubio-Abadal, T. Yefsah, V. Khemani, D. A. Huse, I. Bloch and C. Gross, *Exploring the many-body localization transition in two dimensions*, Science, **352**, 1547 (2016).

30  H. P. Lüschen, P. Bordia, S. Scherg, F. Alet, E. Altman, U. Schneider and I. Bloch, *Observation of Slow Dynamics near the Many-Body Localization Transition in One-Dimensional Quasiperiodic Systems*, Physical Review Letters, **119**, 260401 (2017).

31  R. Islam, R. Ma, P. M. Preiss, M. E. Tai, A. Lukin, M. Rispoli and M. Greiner, *Measuring entanglement entropy in a quantum many-body system*, Nature, **528**, 77 (2015).

32  W. Hofstetter, J. I. Cirac, P. Zoller, E. Demler and M. D. Lukin, *High-Temperature Superfluidity of Fermionic Atoms in Optical Lattices*, Physical Review Letters, **89**, 220407 (2002).

33  C. Becker, P. Soltan-Panahi, J. Kronjäger, S. Dörscher, K. Bongs and K. Sengstock, *Ultracold quantum gases in triangular optical lattices*, New Journal of Physics, **12**, 065025 (2010).

34  P. Soltan-Panahi, J. Struck, P. Hauke, A. Bick, W. Plenkers, G. Meineke, C. Becker, P. Windpassinger, M. Lewenstein and K. Sengstock, *Multi-component quantum gases in spin-dependent hexagonal lattices*, Nature Physics, **7**, 434 (2011).

35  L. Tarruell, D. Greif, T. Uehlinger, G. Jotzu and T. Esslinger, *Creating, moving and merging Dirac points with a Fermi gas in a tunable honeycomb lattice*, Nature, **483**, 302 (2012).

36  G.-B. Jo, J. Guzman, C. K. Thomas, P. Hosur, A. Vishwanath and D. M. Stamper-Kurn, *Ultracold Atoms in a Tunable Optical Kagome Lattice*, Physical Review Letters, **108**, 045305 (2012).

37  J. Sebby-Strabley, M. Anderlini, P. S. Jessen and J. V. Porto, *Lattice of double wells for manipulating pairs of cold atoms*, Physical Review A, **73**, 033605 (2006).

38  S. Fölling, S. Trotzky, P. Cheinet, M. Feld, R. Saers, A. Widera, T. Müller and I. Bloch, *Direct observation of second-order atom tunnelling,* Nature, **448**, 1029 (2007).

39    P. Cheinet, S. Trotzky, M. Feld, U. Schnorrberger, M. Moreno-Cardoner,
      S. Fölling and I. Bloch, *Counting Atoms Using Interaction Blockade in an
      Optical Superlattice*, Physical Review Letters, **101**, 090404 (2008).

40    S. Trotzky, P. Cheinet, S. Fölling, M. Feld, U. Schnorrberger, A. M. Rey,
      A. Polkovnikov, E. A. Demler, M. D. Lukin and I. Bloch, *Time-Resolved
      Observation and Control of Superexchange Interactions with Ultracold
      Atoms in Optical Lattices*, Science, **319**, 295 (2008).

41    D. R. Hofstadter, *Energy levels and wave functions of Bloch electrons in
      rational and irrational magnetic fields*, Physical Review B, **14**, 2239
      (1976).

42    S. M. Girvin, *The Quantum Hall Effect: Novel Excitations and Broken
      Symmetries*, in Topological Aspects of Low Dimensional Systems, Berlin,
      Springer-Verlag, (1998).

43    D. Jaksch and P. Zoller, *Creation of effective magnetic fields in optical*,
      New Journal of Physics, **5**, 56 (2003).

44    M. Aidelsburger, M. Atala, M. Lohse, J. T. Barreiro, B. Paredes and
      I. Bloch, *Realization of the Hofstadter Hamiltonian with Ultracold Atoms
      in Optical Lattices*, Physical Review Letters, **111**, 185301 (2013).

45    H. Miyake, G. A. Siviloglou, C. J. Kennedy, W. C. Burton and
      W. Ketterle, *Realizing the Harper Hamiltonian with Laser-Assisted
      Tunneling in Optical Lattices*, Physical Review Letters, **111**, 185302
      (2013).

46    C. V. Parker, L.-C. Ha and C. Chin, *In situ observation of strongly
      interacting ferromagnetic domains in a shaken optical lattice*, Nature
      Physics, **9**, 769 (2013).

47    H. Lignier, C. Sias, D. Ciampini, Y. Singh, A. Zenesini, O. Morsch and
      E. Arimondo, *Dynamical Control of Matter-Wave Tunneling in Periodic
      Potentials*, Physical Review Letters, **99**, 220403 (2007).

48    X. Li, A. Paramekanti, A. Hemmerich and W. V. Liu, *Proposed formation
      and dynamical signature of a chiral Bose liquid in an optical lattice*,
      Nature Communications, **5**, 3205 (2014).

49    T. Müller, S. Fölling, A. Widera and I. Bloch, *State Preparation and
      Dynamics of Ultracold Atoms in Higher Lattice Orbitals*, Physical Review
      Letters, **99**, 200405 (2007).

50    E. Guardado-Sanchez, P. T. Brown, D. Mitra, T. Devakul, D. A. Huse,
      P. Schauß and W. S. Bakr, *Probing the Quench Dynamics of
      Antiferromagnetic Correlations in a 2D Quantum Ising Spin System*,
      Physical Review X, **8**, 021069 (2018).

51    V. Lienhard, S. d. Léséleuc, D. Barredo, T. Lahaye, A. Browaeys,
      M. Schuler, L.-P. Henry and A. M. Läuchli, *Observing the Space- and
      Time-Dependent Growth of Correlations in Dynamically Tuned Synthetic
      Ising Models with Antiferromagnetic Interactions*, Physical Review X, **8**,
      021070 (2018).

52    H. Labuhn, D. Barredo, S. Ravets, S. d. Léséleuc, T. Macrì, T. Lahaye and
      A. Browaeys, *Tunable two-dimensional arrays of single Rydberg atoms for
      realizing quantum Ising models*, Nature, **534**, 667 (2016).

53    H. Bernien, S. Schwartz, A. Keesling, H. Levine, A. Omran, H. Pichler,
      S. Choi, A. S. Zibrov, M. Endres, M. Greiner, V. Vuletić and M. D. Lukin,
      *Probing many-body dynamics on a 51-atom quantum simulator,* Nature,
      **551**, 579 (2017).

54    J. Zeiher, R. v. Bijnen, P. Schauß, S. Hild, J.-y. Choi, T. Pohl, I. Bloch and
      C. Gross, *Many-body interferometry of a Rydberg-dressed spin lattice*,
      Nature Physics, **12**, 1095 (2016).

55    P. Schauß, J. Zeiher, T. Fukuhara, S. Hild, M. Cheneau, T. Macrì, T. Pohl,
      I. Bloch and C. Gross, *Crystallization in Ising quantum magnets*, Science,
      **347**, 1455 (2015).

56    S. Baier, M. J. Mark, D. Petter, K. Aikawa, L. Chomaz, Z. Cai,
      M. Baranov, P. Zoller and F. Ferlaino, *Extended Bose-Hubbard models
      with ultracold magnetic atoms*, Science, **352**, 201 (2016).

57    A. d. Paz, P. Pedri, A. Sharma, M. Efremov, B. Naylor, O. Gorceix,
      E. Maréchal, L. Vernac and B. Laburthe-Tolra, *Probing spin dynamics
      from the Mott insulating to the superfluid regime in a dipolar lattice gas*,
      Physical Review A, **93**, 021603(R) (2016).

58    S. Lepoutre, J. Schachenmayer, L. Gabardos, B. Zhu, B. Naylor,
      E. Marechal, O. Gorceix, A. M. Rey, L. Vernac and B. Laburthe-Tolra,
      *Exploring out-of-equilibrium quantum magnetism and thermalization in a
      spin-3 many-body dipolar lattice system, arXiv:1803.02628.*

59    M. Greiner, O. Mandel, T. Esslinger, T. W. Hänsch and I. Bloch, *Quantum
      phase transition from a superfluid to a Mott insulator in a gas of ultracold
      atoms*, Nature, **415**, 39 (2002).

60    D. Jaksch, C. Bruder, J. I. Cirac, C. W. Gardiner and P. Zoller, *Cold
      Bosonic Atoms in Optical Lattices*, Physical Review Letters, **81**, 3108
      (1997).

61    M. P. A. Fisher, P. B. Weichman, G. Grinstein and D. S. Fisher, *Boson
      localization and the superfluid-insulator transition*, Physical Review B, **40**,
      546 (1989).

62    M. M. Parish, Chapter 9: The BCS-BEC Crossover, in *Quantum Gas
      Experiments: Exploring Many-Body States*, London, Imperial College
      Press, **179**, (2014).

63    M. Köhl, H. Moritz, T. Stöferle, K. Günter and T. Esslinger, *Fermionic
      Atoms in a Three Dimensional Optical Lattice: Observing Fermi Surfaces,
      Dynamics, and Interactions*, Physical Review Letters, **94**, 080403 (2005).

64    R. Jördens, N. Strohmaier, K. Günter, H. Moritz and T. Esslinger, *A Mott
      insulator of fermionic atoms in an optical lattice*, Nature, **455**, 204 (2008).

65    U. Schneider, L. Hackermüller, S. Will, T. Best, I. Bloch, T. A. Costi,
      R. W. Helmes, D. Rasch and A. Rosch, *Metallic and Insulating Phases of*

*Repulsively Interacting Fermions in a 3D Optical Lattice*, Science, **322**, 1520 (2008).

66  P. M. Duarte, R. A. Hart, T.-L. Yang, X. Liu, T. Paiva, E. Khatami, R. T. Scalettar, N. Trivedi and R. G. Hulet, *Compressibility of a Fermionic Mott Insulator of Ultracold Atoms*, Physical Review Letters, **114**, 070403 (2015).

67  R. Jördens, L. Tarruell, D. Greif, T. Uehlinger, N. Strohmaier, H. Moritz, T. Esslinger, L. D. Leo, C. Kollath, A. Georges, V. Scarola, L. Pollet, E. Burovski, E. Kozik and M. Troyer, *Quantitative Determination of Temperature in the Approach to Magnetic Order of Ultracold Fermions in an Optical Lattice*, Physical Review Letters, **104**, 180401 (2010).

68  D. Greif, M. F. Parsons, A. Mazurenko, C. S. Chiu, S. Blatt, F. Huber, G. Ji and M. Greiner, *Site-resolved imaging of a fermionic Mott insulator*, Science, **351**, 953 (2016).

69  E. Cocchi, L. A. Miller, J. H. Drewes, M. Koschorreck, D. Pertot, F. Brennecke and a. M. Köhl, *Equation of State of the Two-Dimensional Hubbard Model*, Physical Review Letters, **116**, 175301 (2016).

70  L. W. Cheuk, M. A. Nichols, K. R. Lawrence, M. Okan, H. Zhang and M. W. Zwierlein, *Observation of 2D Fermionic Mott Insulators of 40K with Single-Site Resolution*, Physical Review Letters, **116**, 235301 (2016).

71  J. Drewes, E. Cocchi, L. Miller, C. Chan, D. Pertot, F. Brennecke and M. Köhl, T*hermodynamics versus Local Density Fluctuations in the Metal–Mott-Insulator Crossover*, Physical Review Letters, **117**, 135301 (2016).

72  T. Uehlinger, G. Jotzu, M. Messer, D. Greif, W. Hofstetter, U. Bissbort and T. Esslinger, *Artificial Graphene with Tunable Interactions*, Physical Review Letters, **111**, 185307 (2013).

73  J. Imriška, M. Iazzi, L. Wang, E. Gull, D. Greif, T. Uehlinger, G. Jotzu, L. Tarruell, T. Esslinger and M. Troyer, *Thermodynamics and Magnetic Properties of the Anisotropic 3D Hubbard Model*, Physical Review Letters, **112**, 115301 (2014).

74  T. Esslinger, *Fermi-Hubbard Physics with Atoms in an Optical Lattice*, Annual Review of Condensed Matter Physics, **1**, 129 (2010).

75  D. Mitra, P. T. Brown, E. Guardado-Sanchez, S. S. Kondov, T. Devakul, D. A. Huse, P. Schauß and W. S. Bakr, *Quantum gas microscopy of an attractive Fermi–Hubbard system*, Nature Physics, **14**, 173 (2018).

76  P. T. Brown, D. Mitra, E. Guardado-Sanchez, P. Schauß, S. S. Kondov, E. Khatami, T. Paiva, N. Trivedi, D. A. Huse and W. S. Bakr, *Spin-imbalance in a 2D Fermi-Hubbard system*, Science, **357,** 1385 (2017).

77  S. Taie, R. Yamazaki, S. Sugawa and Y. Takahashi, *An SU(6) Mott insulator of an atomic Fermi gas realized by large-spin Pomeranchuk cooling*, Nature Physics, **8**, 825 (2012).

78    C. Hofrichter, L. Riegger, F. Scazza, M. Höfer, D. R. Fernandes, I. Bloch and S. Fölling, *Direct Probing of the Mott Crossover in the SU(N) Fermi–Hubbard model*, Physical Review X, **6**, 021030 (2016).

79    M. Lebrat, P. Grišins, D. Husmann, S. Häusler, L. Corman, T. Giamarchi, J.-P. Brantut and T. Esslinger, *Band and Correlated Insulators of Cold Fermions in a Mesoscopic Lattice*, Physical Review X, **8**, 011053 (2018).

80    D. Greif, T. Uehlinger, G. Jotzu, L. Tarruell and T. Esslinger, *Short-Range Quantum Magnetism of Ultracold Fermions in an Optical Lattice*, Science, **340**, 6138 (2013).

81    R. A. Hart, P. M. Duarte, T.-L. Yang, X. Liu, T. Paiva, E. Khatami, R. T. Scalettar, N. Trivedi, D. A. Huse and R. G. Hulet, *Observation of antiferromagnetic correlations in the Hubbard model with ultracold atoms*, Nature, **519**, 211 (2015).

82    L. W. Cheuk, M. A. Nichols, K. R. Lawrence, M. Okan, H. Zhang, E. Khatami, N. Trivedi, T. Paiva, M. Rigol and M. W. Zwierlein, *Observation of spatial charge and spin correlations in the 2D Fermi–Hubbard model*, Science, **353**, 6305 (2016).

83    M. F. Parsons, A. Mazurenko, C. S. Chiu, G. Ji, D. Greif and M. Greiner, *Site-resolved measurement of the spin-correlation function in the Fermi–Hubbard model*, Science, **353**, 1253 (2016).

84    M. Boll, T. A. Hilker, G. Salomon, A. Omran, J. Nespolo, L. Pollet, I. Bloch and C. Gross, *Spin- and density-resolved microscopy of antiferromagnetic correlations in Fermi–Hubbard chains*, Science, **353**, 1257 (2016).

85    A. Mazurenko, C. S. Chiu, G. Ji, M. F. Parsons, M. Kanász-Nagy, R. Schmidt, F. Grusdt, E. Demler, D. Greif and M. Greiner, *A cold-atom Fermi–Hubbard antiferromagnet*, Nature, **545**, 462 (2017).

86    J. Drewes, L. Miller, E. Cocchi, C. Chan, N. Wurz, M. Gall, D. Pertot, F. Brennecke and M. Köhl, *Antiferromagnetic Correlations in Two-Dimensional Fermionic Mott-Insulating and Metallic Phases*, Physical Review Letters, **118**, 170401 (2017).

87    N. Wurz, C. F. Chan, M. Gall, J. H. Drewes, E. Cocchi, L. A. Miller, D. Pertot, F. Brennecke and M. Köhl, *Coherent manipulation of spin correlations in the Hubbard model*, Physical Review A, **97**, 051602 (2018).

88    E. Cocchi, L. Miller, J. Drewes, C. Chan, D. Pertot, F. Brennecke and M. Köhl, *Measuring Entropy and Short-Range Correlations in the Two-Dimensional Hubbard Model*, Physical Review X, **7**, 031025 (2017).

89    A. W. Sandvik, *Computational Studies of Quantum Spin Systems*, AIP Conference Proceedings, **1297**, 135 (2010).

90    R. Blankenbecler, D. J. Scalapino and R. L. Sugar, *Monte Carlo calculations of coupled boson-fermion systems. I,* Physical Review D, **24**, 2278 (1981).

91    A. Georges, G. Kotliar, W. Krauth and M. J. Rozenberg, *Dynamical mean-field theory of strongly correlated fermion systems and the limit of infinite dimensions*, Reviews of Modern Physics, **68**, 13 (1996).

92    L. Pollet, *Recent developments in quantum Monte Carlo simulations with applications for cold gases*, Reports on Progress in Physics, **75**, 094501 (2012).

93    B. Tang, E. Khatami and M. Rigol, *A short introduction to numerical linked-cluster expansions*, Computer Physics Communications, **184**, 557 (2013).

94    E. J. Mueller, *Review of pseudogaps in strongly interacting Fermi gases*, Reports on Progress in Physics, **80**, 104401 (2017).

95    O. Gunnarsson, M. Calandra and J. E. Han, *Colloquium: Saturation of electrical resistivity*, Reviews of Modern Physics, **75**, 1085 (2003).

96    C.-L. Hung, X. Zhang, N. Gemelke and C. Chin, *Slow Mass Transport and Statistical Evolution of an Atomic Gas across the Superfluid–Mott-Insulator Transition*, Physical Review Letters, 104, 160403 (2010).

97    D. C. M. a. B. DeMarco, *Cooling in strongly correlated optical lattices: prospects and challenges*, Reports on Progress in Physics, **74**, 054401 (2011).

## 9. Real time control of Hamiltonian of a 2D quantum gas

Cheng Chin

*University of Chicago*

An intriguing development in modern atomic physics research comes from the realization that at sufficiently low temperatures, the Hamiltonian of quantum gases can be easily engineered and programmed based on electromagnetic fields. The current state-of-the-art offers an intriguing prospect to gain full control of the Hamiltonian of a quantum many-body system to simulate novel quantum phenomena and implement quantum information processing.

Here we consider a generic Hamiltonian $H$ of atoms confined in an optical lattice. It includes kinetic energy $K$, interaction energy $U$ and potential energy $V$, given by

$$H = K + U + V = \sum_{<j,k>} (t_{j,k} a_j^+ a_k + H.c.) + \sum_j U_j \frac{n_j(n_{j-1})}{2} - \sum_j \mu_j n_j,$$

where the summation goes over all sites in a 2D lattice, $t_{j,k}$ is the locally defined tunneling between neighboring sites, $U_j$ is the locally defined interaction, and $\mu_j$ is the locally defined potential energy.

Full control of the Hamiltonian, including all three terms, is based on a powerful toolbox that we will develop. In the following, we will describe the ideas and how one employs and integrates them to gain control of the system on single atom level and in real time.

1. *Control of lattice potential: Holographic projection based on spatial light modulation*

The state-of-art in quantum control of cold atoms involves spatial light modulators, a modern tool to engineer arbitrary light pattern with high spatial and temporal resolution. An example is the Digital Micromirror Device (DMD), which contains millions of metallic mirrors. Each mirror is few microns in size that can be independently and rapidly switched. An application of the DMD is to holographically project an arbitrary lattice pattern to confine atoms. Such devices have been widely employed in recent cold-atom experiments to generate quantum gas in a square well, optical lattice, ring-shape, and even random potentials. Here we summarize some of the results from our laboratory at the University of Chicago and ways to further improve its performance.

A holographically formed optical lattice is built site by site. It can be constructed based on various approaches: diffractive optical elements, high resolution liquid crystal modulators, or DMDs. All of the approaches generate very flexible lattice geometry, for example, square, triangular, honeycomb and

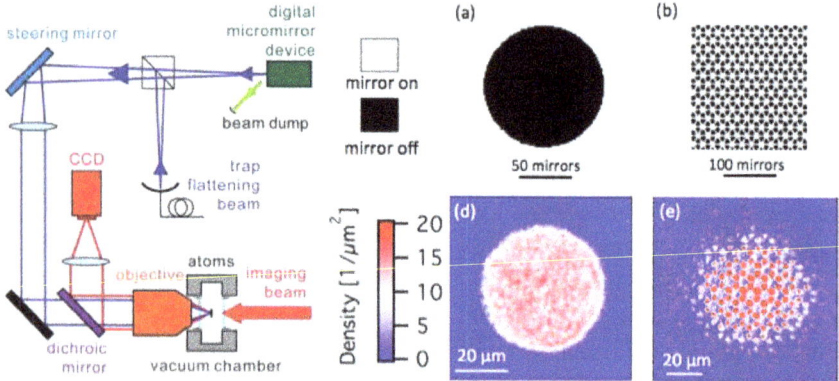

Fig. 21. (Left panel) The optical potential patterning realized based on a digital micromirror device (DMD). The dichroic mirror combines imaging and patterned beams into the high-resolution objective. (Right panel) (a) and (d) DMD pattern that confines $10^4$ cesium atoms in a 2D flat-bottom trap. (b) and (e) DMD holographically generates a honeycomb optical lattice.

even quasi-crystal lattices. DMDs, in particular, provide great flexibility superior stability, and reproducibility as compared to other devices. It, however, also suffers from weak optical interference and requires fast electronics to program all the mirrors.

At the University of Chicago, we employ a DMD unit to reflect a blue-detuned laser at 532 nm. A projected resolution of 465 nm is achieved, consistent with the diffraction limit and lattice spacing. We project the pattern onto atoms using a microscope objective lens designed for imaging and optical tweezers. The DMD provides us with the flexibility to generate an arbitrary optical potential. A circularly flat potential and a honeycomb lattice are shown in Fig. 21.

As an example, 10s to 100s of optical tweezer beams can be programmed with a DMD to confine and drag atoms on a 2D plane to different locations. This generalizes the recent work on the one dimensional control of 51 atoms based on optical tweezers.[1] One application of the atom patterning is to develop an algorithm to pack randomly distributed atoms in 2D lattices into a Mott insulating domain with exactly 1 atom per lattice site, illustrated in Fig. 22. With high fidelity control of the tweezers and optical cooling, the entropy of the system can be greatly suppressed, and we expect the sample can enter the quantum regime (Mott insulator) directly without significant atom loss. By reducing the lattice potential, a Mott insulating sample can also be converted into a regular atomic superfluid.

## 2. Optical control of atomic interactions

Control of atomic interactions presents a new way to watch the emergence of strongly correlated systems from a weakly interacting one. Such control by an external radiation field can be realized in stable Bose–Einstein condensates,[2] and

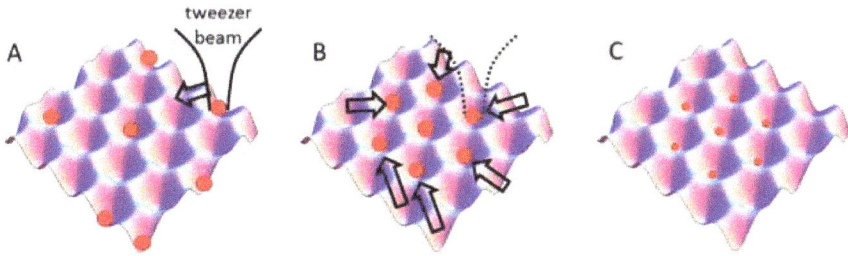

Fig. 22. Patterning atoms using optical tweezers. (A) After MOT, atoms randomly populate a deep optical lattice. (B) Using optical tweezers, atoms can be rearranged to form a desired pattern. (C) After optically cooling the atoms into the ground state of the lattice sites, the sample forms a 2D Mott insulator.

is called optical Feshbach resonance (OFR), see Fig. 23. OFR is significantly more versatile than the magnetic counterpart since optical control can be implemented with very high speed and high special resolution. In recent experiments, switching atomic interaction at the time scale of 10 ns and a local control of atomic interactions within a few-μm length scale led to interesting quantum dynamics. These results cannot be easily realized with magnetic field control.

We note that improving the performance of OFR with even higher spatial resolution and temporal speed is straightforward. In time, one can employ optical modulation at 10 GHz to realize switching speed of atomic interaction in < 1 ns. In space, by directly sending the light through the optical objective, a local control of the scattering length at the length scale of one lattice constant (500 nm) can be realized.

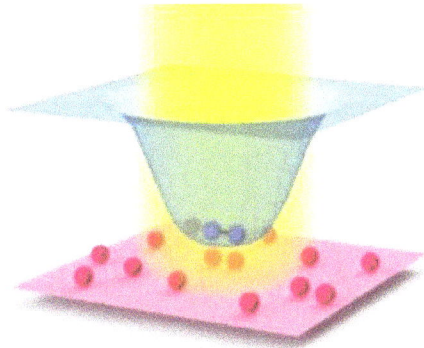

Fig. 23. Optical control of atomic interactions in quantum gases has been a long-sought goal for 20 years. We developed and implemented a generic scheme for optical Feshbach resonances, which yields long quantum gas lifetimes and a negligible parasitic radiation force. The fast and local control of interactions leads to intriguing quantum dynamics in new regimes.[2]

### 3. *Synthetizing gauge field by lattice and magnetic field modulation*

Synthetic gauge fields refer to artificial gauge potential that affect the dynamics of neutral atoms in the same way as electromagnetic fields on charged particles in solid state or in high energy physics.[3,4] In lattice systems, this gauge field enters the Hamiltonian as a complex tunneling term between lattice sites, which manifests as a "Peierls phase" when atoms tunnel to their neighboring sites.[5]

The dynamical feedback between matter and gauge fields is a new frontier for cold atoms to explore even broader physics ranging from quantum material design[6] to quantum chromodynamics.[7] Realization of dynamical gauge fields based on cold atom systems has been extensively discussed.[8-12] One crucial step is to realize the gauge field that depends on the temporal modulation of atomic interactions.[13]

An example of the interaction-induced gauge field is illustrated in Fig. 24. Within one Floquet period of lattice modulation, the micromotion of the atomic wavefunction shows both periodic displacement and density modulation. In particular, the density oscillates at $\pm 90°$ relative to the lattice shaking for atoms with $+/-$ quasi-momentum $\boldsymbol{k}$.

One may break the inversion or the time reversal symmetry $\boldsymbol{k} \to -\boldsymbol{k}$ by modulating the atomic interactions in-phase or out-of-phase with the density modulation. An in-phase modulation of the scattering length with the density, for example, yields a larger than average interaction: $\overline{g(t)n(t)} > \overline{g(t)}\,\overline{n(t)}$. Thus one expects that a simultaneous modulation of both lattice and scattering length will result in an effective synthetic gauge potential in the form of $H \approx h(k;s) +$

a)

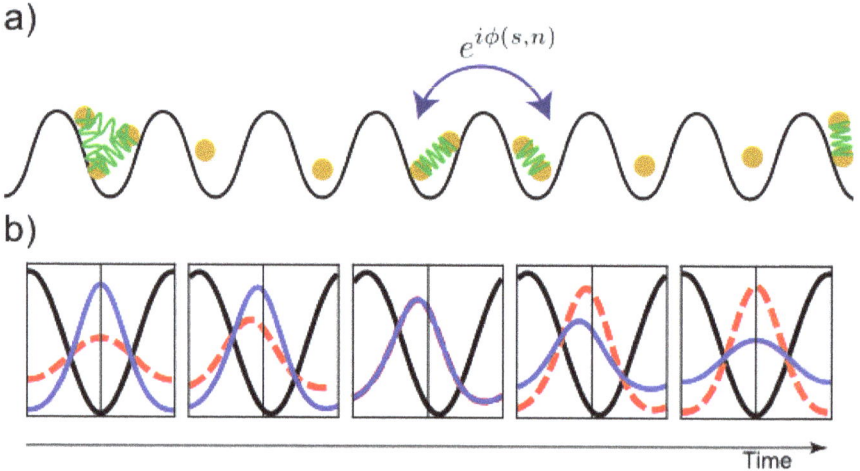

b)

Time

Fig. 24. a) Illustrates the non-trivial phase shift, due to shaking and interaction, while atoms hop from site to site in the lattices; b) Demonstrates the micro-motion of atomic wavefunctions at two momenta $+\boldsymbol{k}$ and $-\boldsymbol{k}$ (Blue and Red) within a Floquet period. Due to the lattice shaking (black), we expect that a synchronous modulation of interaction can thus break the inversion symmetry $\boldsymbol{k} \to -\boldsymbol{k}$.

$f(k)g_{AC}s\bar{n}$, where $h(k; s)$ is the single-particle dispersion and $f(k)g_{AC}s\bar{n}$ is the effective gauge field that depends on the atomic density $n$, shaking amplitude $s$, and the amplitude of the interaction modulation $g_{AC}$. The symmetry breaking property of the gauge field demands that $f(k) = -f(-k)$. Such coupling between momentum states and atomic density can be described as based on an interaction-dependent synthetic gauge field and can lead to intriguing anyonic excitations.[14]

## References

1  Hannes Bernien, Sylvain Schwartz, Alexander Keesling, Harry Levine, Ahmed Omran, Hannes Pichler, Soonwon Choi, Alexander S. Zibrov, Manuel Endres, Markus Greiner, Vladan Vuletić, Mikhail D. Lukin, *Probing many-body dynamics on a 51-atom quantum simulator,* Nature **551**, 579–584 (2017).

2  Logan W. Clark, Li-Chung Ha, Chen-Yu Xu, and Cheng Chin, *Quantum Dynamics with Spatiotemporal Control of Interactions in a Stable Bose–Einstein Condensate,* Phys. Rev. Lett. **115**, 155301 (2015).

3  Y.-J. Lin, K. J. García, and I. B. Spielman, *A spin-orbit coupled Bose–Einstein condensate,* Nature **471**, 83 (2011).

4  Y.-J. Lin, R. L. Compton, K. J. Garcia, J. V. Porto, and I. B. Spielman, *Synthetic magnetic fields for ultracold neutral atoms,* Nature **462**, 628 (2009).

5  K. J. Garcia, L. J. LeBlanc, R. A. Williams, M. C. Beeler, A. R. Perry, and I. B. Spielman, *Peierls substitution in an engineered lattice potential,* Phys. Rev. Lett. **108**, 225303 (2012).

6  M. Levin and X. G. Wen, *Colloquium: Photons and electrons as emergent phenomena,* Rev. Mod. Phys. **77**, 871 (2005).

7  J. Kogut, *The lattice gauge theory approach to quantum chromodynamics,* Rev. Mod. Phys. **55**, 775 (1983).

8  U. J. Wiese, *Ultracold quantum gases and lattice systems: quantum simulation of lattice gauge theories,* Ann. Phys. (N.Y.) **525**, 777 (2013).

9  J. I. Cirac, P. Maraner, and J. K. Pachos, *Cold Atom Simulation of Interacting Relativistic Quantum Field Theories,* Phys. Rev. Lett.**105**, 190403 (2010).

10  E. Zohar, J. I. Cirac, B. Reznik, *Cold-Atom Quantum Simulator for SU(2) Yang–Mills Lattice Gauge Theory,* Phys. Rev. Lett. **110**, 125304 (2013).

11  Eliot Kapit and Erich Mueller, *Optical-lattice Hamiltonians for relativistic quantum electrodynamics,* Phys. Rev. A **83**, 033625 (2011).

12  Erez Zohar, J. Ignacio Cirac, and Benni Reznik, *Simulating Compact Quantum Electrodynamics with Ultracold Atoms: Probing Confinement and Nonperturbative Effects,* Phys. Rev. Lett. **109**, 125302 (2012).

13  S. Greschner, G. Sun, D. Poletti, and L. Santos, *Density-Dependent Synthetic Gauge Fields Using Periodically Modulated Interactions,* Phys. Rev. Lett. **113**, 215303 (2014).

14    L. W. Clark, B. M. Anderson, L. Feng, K. Levin, C. Chin, *Observation of Density-Dependent Gauge Fields in a Bose–Einstein Condensate Based on Micromotion Control in a Shaken Two-Dimensional Lattice*, Phys. Rev. Lett. **121**, 030402 (2018).

## 10. Trapped ions

Norbert Linke

*University of Maryland*

Since there cannot be an electric field minimum in free space, (Earnshaw's theorem[1]) ion traps rely on dynamic confinement[2] using oscillating radio-frequency fields combined with static potentials to confine charged particles in all three directions. The most common geometry is the linear Paul trap, where a strong radial RF-quadrupole field generated by four elongated electrodes, and a weaker axial DC potential leads to a linear chain of ions along the field-free center line of the quadrupole.[3] The DC potential can either be generated by endcaps or segments on the RF-grounded electrode pair as shown in Fig. 25.

This geometry can also be folded into a single plane generating a trapping potential above the surface.[4] State-of-the-art surface traps (see Fig. 26) include the ability to shuttle ions along the chip and reorder them using junctions that connect separate trap arms.[5] Additionally, more sophisticated control elements such as microwave-current carrying wires and microwave resonators[6] as well as optical waveguides that can deliver signals directly to the ions in an integrated fashion.[7]

There are two types of spin-½ systems (or qubits) used in trapped ions. One is optical qubits where one state is in the ground level and one is in a low-lying D-level. Secondly, there are ground level qubits that use pairs of states in the Zeeman-split or hyperfine-split ground state. The latter type has been the focus in our group using $^{171}$Yb$^+$ ions. The ground level has a pair of states that are first-order insensitive to magnetic field changes, leading to a coherence time in excess of a second without any magnetic shielding or active field stabilization. Using spin-echo techniques, coherence times of up to 10 mins. have been measured.[8] State initialization by optical pumping and readout by state-dependence

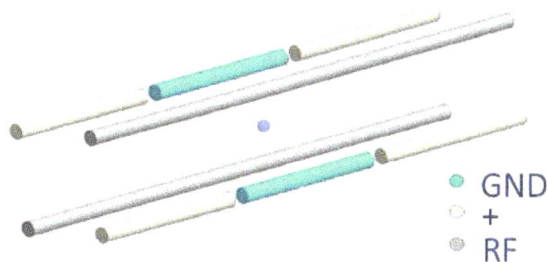

Fig. 25. A segmented linear Paul trap with a trapped ion indicated in the center. The RF electrodes generate an oscillating radial quadrupole field for dynamic confinement while the segmented electrodes generate axial static confinement.

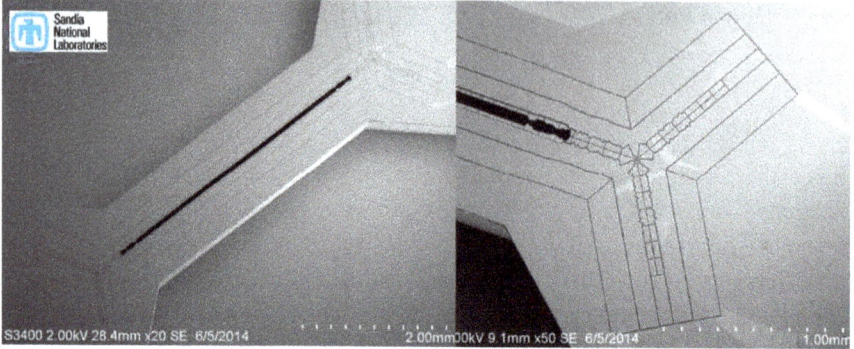

Fig. 26. Electron microscope image of the HOA-2 (High Optical Access Trap 2) trap made by Sandia National Labs. The gold surface is patterned into electrodes for shuttling ions along the central section while a 3-way junction can be used to reorder ion crystals.

fluorescence detection can be done with high fidelity.[9] Qubit transitions can be driven by microwave radiation[10] or a pair of Raman beams[11] for ground level qubit, as well as direct laser transitions on the quadrupole transition for optical qubits.[12]

To create specific Hamiltonians, there are different ways to apply addressed operation on individual ions. One is by using an addressed Stark-shift beam that effects a z-rotation[13] while the other is an individually addressed Raman beam scheme.[11] To create two-qubit interactions, the ions' quantized motion is used as a type of bus mode as mentioned above. By setting the laser frequency (or beatnote frequency in the case of a Raman scheme) off resonance such that the detuning corresponds to the secular trap frequency we can drive so-called blue or red sideband transitions in which the spin state is changed while simultaneously adding or removing a phonon from or to the motional mode, respectively (see Fig. 27).

To evoke an effective spin-spin or two-qubit Hamiltonian, two laser frequencies near the red and the blue sidebands are applied at the same time.[14-16] This is illustrated in Fig. 27. When applied globally to a chain of ions, the resulting system follows a long-range Ising model with effective transverse field $B$ as follows:

$$H_{eff} = \sum_{i<j} J_{ij}\, \sigma_i^x \sigma_j^x + \frac{B}{2} \sum_i \sigma_i^z ,$$

where the coupling constants $J_{ij}$ given by

$$J_{ij} \approx \frac{J_0}{|i-j|^\alpha}$$

falls off with distance raised to an exponent $\alpha$ that can be changed between 0 and 3 using laser detuning.

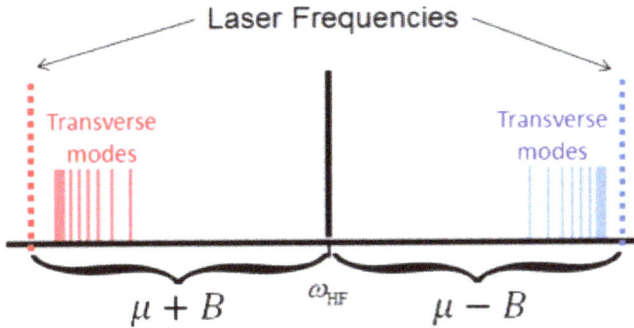

Fig. 27. Application of two laser frequencies near the motional modes (detuning μ from carrier transition $\omega_{HF}$). The imbalance of the detuning adds an effective $B$-field to the realized Ising system (see text).

A measurement of this relationship is shown in Ref. [17]. This mechanism can be used to study a range of different physical phenomena such as many-body localization,[17] quenches,[18] pre-thermalization,[19] XY- and Ising model simulations,[20] including the observation of phase transitions,[21] the demonstration of discrete time crystals,[21] and many others.

These examples show that trapped ions are a versatile and powerful platform for quantum simulation (as well as quantum information processing, see below). One of the major challenges facing the field going forward is to increase the system size so that simulations can go beyond the capabilities of classical computers. I will now discuss two different efforts currently being pursued in this direction. One of the issues in adding more and more ions to the system is that the lifetime of an ion chain is limited by the background gas pressure in the vacuum system. Background gas molecules, primarily hydrogen, can collide with the chain. There are two types of collisions, elastic and inelastic. In the first case, energy is transferred, leading to motional excitation. Depending on the collision parameters, this motion can be removed by laser re-cooling with the possibility that the ions swap position within the chain.[22] It can also lead to an excitation large enough that the ions move far enough from the trap center that the harmonic approximation for the trap potential no longer holds. In this case, the dynamic RF-fields that provide confinement near the RF null, can now drive additional energy into the system until the ions can escape the trap. In an inelastic collision, one of the ions forms a hydride molecule and becomes transparent to the wavelengths used for the atomic ion and can no longer be addressed.

In order to reduce the pressure in the chamber, one can employ a cryogenic setup where the trap and the surrounding chamber are held at 4 K or below. At this temperature, residual gas molecules have a high chance of being adsorbed on surfaces, which lowers the pressure. Additionally, any collisions that do happen

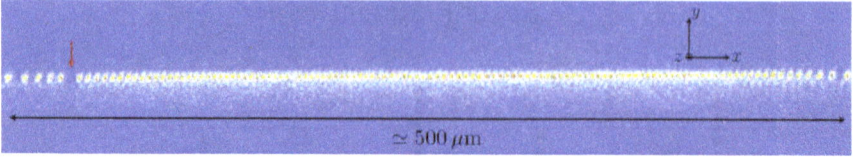

Fig. 28. A stable chain of over 120 ions in a cryogenically cooled ion trap.[23] Red arrow indicates a dark ion, the position of which was observed to be stable over several hours, indicating that no high-energy background gas collisions are occurring (see text).

will be at lower energies, thereby avoiding ion ejecting or reordering events. Such a system has recently demonstrated a stable chain of >120 ions.[23] (See Fig. 28)

The second effort to increase the scale and simulating power of trapped ion systems is to create two-dimensional ion arrays. Having a structure that exhibits native interactions in two dimensions will allow for simulations to probe along new lines of inquiry in quantum-many-body systems such spin glasses and other exotic phases of matter,[24] high-temperature superconductivity,[25] and geometric frustration.[26]

There are several strategies currently being used experimentally for realizing such an arrangement. The most advanced is the use of Penning traps, where the transversal confinement is achieved not by an RF field but by a strong magnetic field.[27] One issue with this approach is that the entire crystal of charged particles rotates in the trap at 100 s of kHz, making individual addressing of single ions challenging. Therefore, two-dimensional ion crystals are being pursued with the RF (Paul) traps described above. One effort aims to achieve a strong axial confinement which pushes the ion configuration from a linear arrangement into a plane in the radial direction.[28] This will allow for axial laser beam addressing (see Fig. 29). An alternative strategy is a multi-trap surface electrode chip that allows for a pattern of individual traps, each holding a single ion. This gives an additional degree of control over the trap parameters at each site as well as the distance between sites when designing the chip. One such device recently described in Ref. [29] in shown in Fig. 30. Since each ion is trapped in the center of its own RF

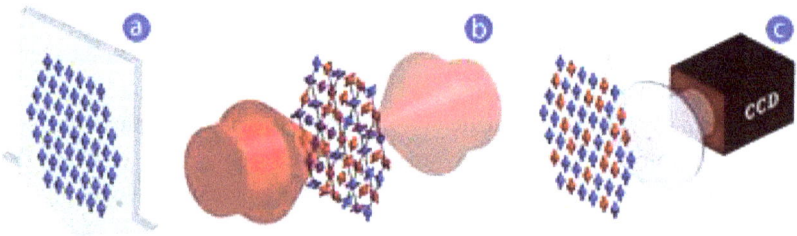

Fig. 29. Increasing the axial confinement in a Paul trap to be larger than the radial leads to a "pancake" type 2D ion crystal. The figure shows global addressing (a), individual addressing (b), and readout by imaging (c). Figure from Richerme group, Indiana.[28]

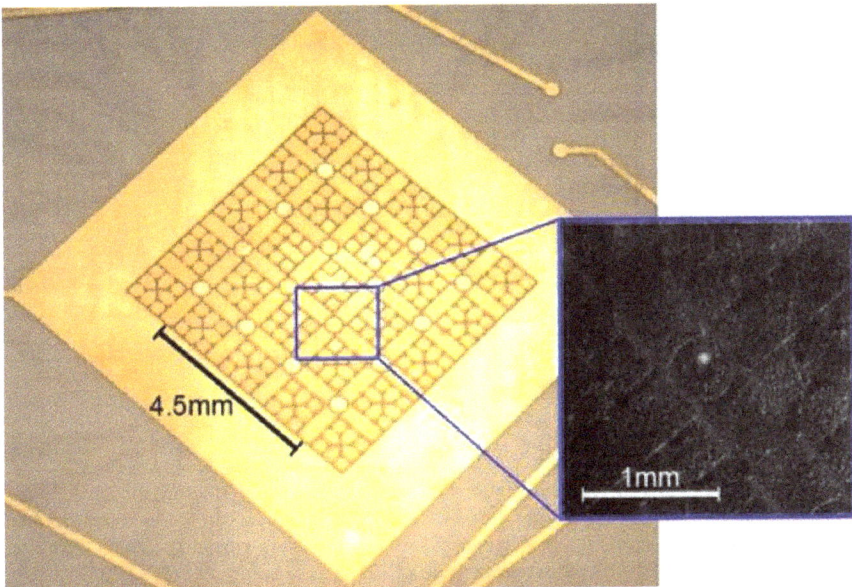

Fig. 30. Trap array "Folsom" with the inset showing a single trapped ion (Blatt group, Innsbruck). An array of trap zones with reconfigurable positions is patterned on the surface. See Ref. [29] for details.

potential, the micro-motion induced by the trapping field can be compensated individually which is not the case for a large 2D crystal in a single trap where most of the ions will be confined off-center — though an oblate ion trap geometry has been proposed to mitigate this issue.[30]

Finally, there is one more way that trapped ions can be used to simulate quantum mechanical models. Trapped ions are among the most advanced machines for creating universal gate-based quantum computers.[31,32] Using the Molmer–Sorensen scheme[14] mentioned above, entangled gates between a subset of trapped ion qubits can be implemented. Using a compiler, these gates can be assembled into arbitrary circuits. Hamiltonians of interest can be converted to qubit circuits using the Jordan–Wigner,[33] Bravyi–Kitaev,[34] or a first-quantization mapping,[35] and simulated approximately by Trotterization. In a recent simulation experiment of this kind, our group has extracted the second Renyi entropy which is a measure of the entanglement between two subsystems.[35] As for the model Hamiltonian, we used the Fermi–Hubbard model,[36] which is seen as a widely applicable model for complex phenomena in condensed matter physics such as high-temperature superconductivity, and hence a key application for quantum computers.[37] Our circuit involves simulating two copies of a two-site Hamiltonian (see Fig. 31). Despite the simplicity of the model, the extraction of the Renyi entropy shows promise for scaled-up versions of this processor that can tackle problems intractable on classical machines. This effort constitutes the deepest

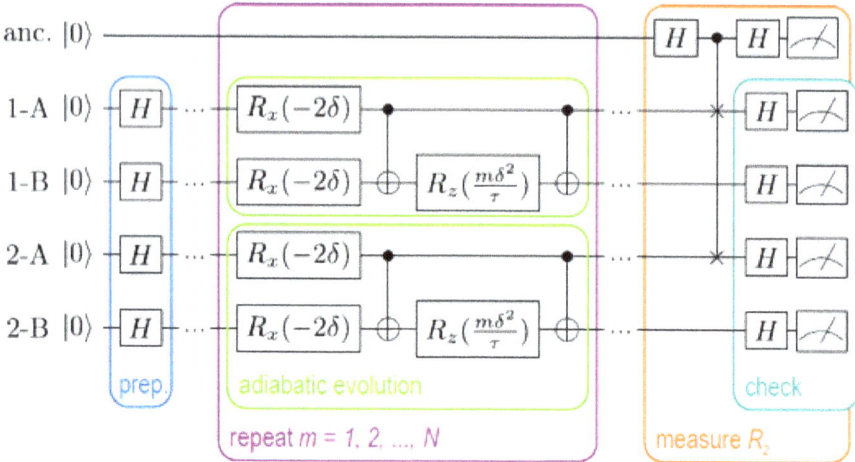

Fig. 31. Circuit to measure the Renyi entropy R2 of a two-site Fermi–Hubbard model using two systems (1 and 2), each with two subsystems (A and B) as implemented on an ion trap quantum computer. The circuit shows the preparation of the non-interacting ground state, the Trotterized adiabatic evolution, and the R2 measurement using a C-Swap gate. For details see Ref. [35].

quantum circuit successfully implemented to date. Efforts to scale up this system by increasing the number of ions controlled, as well as by connecting distant modules are underway.[38]

## References

1   D. J. Griths, *Introduction to Electrodynamics*, Prentice Hall, 3[ed.] (1999).
2   W. Paul, *Electromagnetic traps for charged and neutral particles*, Rev. Mod. Phys. **62**, 531–540 Jul (1990).
3   D. J. Wineland, C. Monroe, W. M. Itano, D. Leibfried, B. E. King, and D. M. Meekhof, *Experimental issues in coherent quantum-state manipulation of trapped atomic ion*, Journal of Research of the National Institute of Standards and Technology **103** 3, 259–328 (1998).
4   S. Seidelin, J. Chiaverini, R. Reichle, J. J. Bollinger, D. Leibfried, J. Britton, J. H. Wesenberg, R. B. Blakestad, R. J. Epstein, D. B. Hume, W. M. Itano, J. D. Jost, C. Langer, R. Ozeri, N. Shiga, and D. J. Wineland, *Microfabricated surface-electrode ion trap for scalable quantum information processing*, Phys. Rev. Lett. **96**, 253003 Jun (2006).
5   J. Amini, H. Uys, J. H. Wesenberg, S. Seidelin, J. Britton, J. J. Bollinger, D. Leibfried, C. Ospelkaus, A. P. VanDevender, and D. J. Wineland, *Toward scalable ion traps for quantum information processing*, New Journal of Physics **12**, Mar (2010).
6   D. T. C. Allcock, T. P. Harty, C. J. Ballance, B. C. Keitch, N. M. Linke, D. N. Stacey, and D. M. Lucas, *A microfabricated ion trap with integrated microwave circuitry*, Applied Physics Letters **102**, 4, 044103 (2013).

7   K. K. Mehta, C. D. Bruzewicz, R. McConnell, R. J. Ram, J. M. Sage, and J. Chiaverini, *Integrated optical addressing of an ion qubit*, Nature Nanotechnology **11**, 1066 EP –, Aug (2016).

8   Y. Wang, M. Um, J. Zhang, S. An, M. Lyu, J.-N. Zhang, L.-M. Duan, D. Yum, and K. Kim, *Single-qubit quantum memory exceeding ten-minute coherence time*, Nature Photonics **11**, 10, 646–650 (2017).

9   S. Olmschenk, K. C. Younge, D. L. Moehring, D. N. Matsukevich, P. Maunz, and C. Monroe, *Manipulation and detection of a trapped* $Yb^+$ *hyperfine qubit*, Phys. Rev. A **76**, 052314 Nov (2007).

10  C. Piltz, T. Sriarunothai, S. S. Ivanov, S. W"olk, and C. Wunderlich, *Versatile microwave-driven trapped ion spin system for quantum information processing*, Science Advances **2**, 7 (2016).

11  C. Monroe and J. Kim, *Scaling the ion trap quantum processor* Science **339**, 6124, 1164–1169 (2013).

12  J. Benhelm, G. Kirchmair, C. F. Roos, and R. Blatt, *Towards faulttolerant quantum computing with trapped ions*, Nature Physics **4**, 463 EP –, Apr (2008).

13  A. C. Lee, J. Smith, P. Richerme, B. Neyenhuis, P. W. Hess, J. Zhang, and C. Monroe, *Engineering large stark shifts for control of individual clock state qubits*, Phys. Rev. A **94**, 042308 Oct (2016).

14  K. Mølmer and A. Sørensen, *Multiparticle entanglement of hot trapped ions*, Phys. Rev. Lett. **82**, 1835–1838 Mar (1999).

15  D. Porras and J. I. Cirac, *Effective quantum spin systems with trapped ions*, Phys. Rev. Lett. **92**, 207901 May (2004).

16  K. Kim, M.-S. Chang, R. Islam, S. Korenblit, L.-M. Duan, and C. Monroe, *Entanglement and tunable spin-spin couplings between trapped ions using multiple transverse modes*, Phys. Rev. Lett. **103**, 120502 Sep (2009).

17  P. W. Hess, P. Becker, H. B. Kaplan, A. Kyprianidis, A. C. Lee, B. Neyenhuis, G. Pagano, P. Richerme, C. Senko, J. Smith, W. L. Tan, J. Zhang, and C. Monroe, *Non-thermalization in trapped atomic ion spin chains*, Phil. Trans. R. Soc. A **375**, 20170107 (2017).

18  J. Schachenmayer, B. P. Lanyon, C. F. Roos, and A. J. Daley, *Entanglement growth in quench dynamics with variable range interactions*, Phys. Rev. X **3**, 031015 Sep (2013).

19  B. Neyenhuis, J. Zhang, P. W. Hess, J. Smith, A. C. Lee, P. Richerme, Z.-X. Gong, A. V. Gorshkov, and C. Monroe, *Observation of prethermalization in long-range interacting spin chains*, Science Advances **3**, 8 (2017).

20  T. Brydges, A. Elben, P. Jurcevic, B. Vermersch, C. Maier, B. P. Lanyon, P. Zoller, R. Blatt, and C. F. Roos, *Probing entanglement entropy via randomized measurements*, arxiv:1806.05747 (2018).

21  J. Zhang, P. W. Hess, A. Kyprianidis, P. Becker, A. Lee, J. Smith, G. Pagano, I.-D. Potirniche, A. C. Potter, A. Vishwanath, N. Y. Yao, and

C. Monroe, *Observation of a discrete time crystal*, Nature **543**, 217 EP –, Mar (2017).

22    P. Bowe, L. Hornekær, C. Brodersen, M. Drewsen, J. S. Hangst, and J. P. Schiffer, *Sympathetic crystallization of trapped ions*, Phys. Rev. Lett. **82**, 2071–2074 Mar (1999).

23    G. Pagano, P. W. Hess, H. B. Kaplan, W. L. Tan, P. Richerme, P. Becker, A. Kyprianidis, J. Zhang, E. Birckelbaw, M. R. Hernandez, Y. Wu, and C. Monroe, *Cryogenic trapped-ion system for large scale quantum simulation*, arXiv:1802.03118 (2018).

24    M. P. A. F. F. Alet, A. M. Walczak, *Exotic quantum phases and phase transitions in correlated matter*, Physics A **369**, 122 (2006).

25    P. W. Anderson, *The resonating valence bond state in* $La_2CuO_4$ *and superconductivity*, Science **235**, 4793, 1196–1198 (1987).

26    R. Moessner and S. L. Sondhi, *Ising models of quantum frustration*, Phys. Rev. B **63**, 224401 May (2001).

27    J. G. Bohnet, B. C. Sawyer, J. W. Britton, M. L. Wall, A. M. Rey, M. FossFeig, and J. J. Bollinger, *Quantum spin dynamics and entanglement generation with hundreds of trapped ions*, Science **352**, 6291, 12971301 (2016).

28    P. Richerme, *Two-dimensional ion crystals in radio-frequency traps for quantum simulation*, Phys. Rev. A **94**, 032320 Sep (2016).

29    M. Kumph, P. Holz, K. Langer, M. Meraner, M. Niedermayr, M. Brownnutt, and R. Blatt, *Operation of a planar-electrode ion-trap array with adjustable rf electrodes*, New Journal of Physics **18**, Feb (2016).

30    B. Yoshimura, M. Stork, D. Dadic, W. C. Campbell, and J. K. Freericks, *Creation of two-dimensional coulomb crystals of ions in oblate paul traps for quantum simulations*, EPJ Quantum Technology **2**, (2015).

31    S. Debnath, N. M. Linke, C. Figgatt, K. A. Landsman, K. Wright, and C. Monroe, *Demonstration of a small programmable quantum computer module using atomic qubits*, Nature **536**, 63–66 Aug (2016).

32    T. Monz, D. Nigg, E. A. Martinez, M. F. Brandl, P. Schindler, R. Rines, S. X. Wang, I. L. Chuang, and R. Blatt, *Realization of a scalable Shor algorithm*, Science **351**, 6277, 1068–1070 (2016).

33    P. Jordan and E. Wigner, *¨Uber das paulische ¨aquivalenzverbot*, Zeitschrift f¨ur Physik **47**, 631–651 Sep (1928).

34    S. B. Bravyi and A. Y. Kitaev, *Fermionic quantum computation*, Annals of Physics **298**, 1, 210–226 (2002).

35    N. M. Linke, S. Johri, C. Figgatt, K. A. Landsman, A. Y. Matsuura, and C. Monroe, *Measuring the renyi entropy of a two-site Fermi–Hubbard model on a trapped ion quantum computer*, arxiv:1712.08581 Dec (2017).

36    J. Hubbard, *Electron correlations in narrow energy bands*, Proc. R. Soc. **276**, 238–257 Nov (1963).

37  D. Wecker, M. B. Hastings, N. Wiebe, B. K. Clark, C. Nayak, and
    M. Troyer, *Solving strongly correlated electron models on a quantum
    computer*, Phys. Rev. A **92**, 062318 Dec (2015).

38  C. Monroe, R. Raussendorf, A. Ruthven, K. R. Brown, P. Maunz,
    L.-M. Duan, and J. Kim, *Large-scale modular quantum-computer
    architecture with atomic memory and photonic interconnects*, Phys. Rev. A
    **89**, 022317 Feb (2014).

## 11. Quantum simulation using quantum dot arrays

Neil Zimmerman

*National Institute of Standards and Technology*

### Introduction

Here, we will consider a variety of predicted and measured results in semiconducting quantum dot arrays, including but not limited to analog quantum simulations.

We first ask: What is a "semiconducting quantum dot?"[1] Briefly, this construction is an isolated puddle of charge made in a semiconducting material. In this context, *semiconducting* is important because it means that we can often control either the density of carriers and/or the electrical conductivity of the barriers that separate the isolated puddle of charge from the conducting leads; this is in contrast to the typical metallic or superconducting devices, where such control of density and barrier conductivity is less feasible. In addition, *semiconducting* is important because it means that we have orders of magnitude fewer carriers (electrons or holes) in the quantum dot, again compared to metallic or superconducting devices — this has import in a variety of ways, such as making the carrier effective mass and the orbital energy spacing much larger than in metals.

Semiconducting quantum dots can be made in a variety of ways, ranging from surface-gated bulk or mesa-etched Si and GeSi wafers to similar GaAs devices to self-assembled nano crystals[1] to nanowires;[2] here, we use the word "gate" to mean a highly-conducting wire on the surface of the semiconductor, which conveys a voltage used to turn on or off conduction in the underlying semiconductor. In general, these devices have a region of relatively high carrier density (the isolated puddle or quantum dot) induced by an "accumulation" gate, and regions of much lower carrier density (the barriers) induced by "depletion" gates in planar devices or by chemical junctions in nanowire devices. The typical lateral (in the plane of the wafer) length scale of these devices is hundreds of nanometers or smaller.

Measurements in these devices are typically done at cryogenic temperatures of 4 K or below, in order to avoid thermal smearing. In this size and temperature range, these devices often exhibit a Coulomb blockade effect, where the effects of single electrons become a dominant part of the electrical behavior. As an example, the Coulomb blockade can result in the most sensitive charge electrometers, useful for nanometer-scale sensitive measurements of charge motion.

In addition to the uses considered in detail in this section, other non-photonic uses of quantum dot devices include: (i) Accurate standards of electrical current,

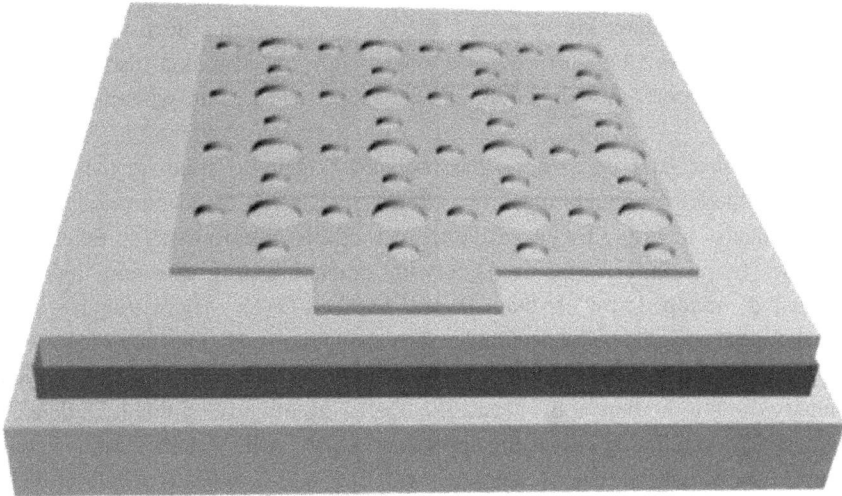

Fig. 32. A proposal for a two-dimensional quantum dot array formed by a top metal electrode on top of planar semiconductor such as GaAs or Si. (From Ref. [7])

based on pumping one electron at a time;[3] (ii) quantum coherent manipulation of single electrons, recently mostly spin but also including charge;[4] and (iii) single-electron classical computing including quantum cellular automata.[5,6]

**Theoretical proposals and results**

There have been a number of proposals/predictions for analog quantum simulations (AQS) in semiconductor quantum dot arrays, dating back about 20 years. We will discuss in detail several of these proposals.

An early proposal,[7] which emphasized the concept of a "dedicated simulator" before the widespread use of the term AQS, suggested using a two-dimensional quantum dot array to simulate the "Hubbard-like" Hamiltonian that was believed to underlie the physics of the Cu-O plane in high temperature superconductors. An early proposal for making a quantum dot array in a simple way is shown in Fig. 32; although suggested to be done in GaAs, it could equally well be done in Si or any other planar semiconductor. In this case, the gate was used to generate an array by repelling the electrons from underneath the shown electrode, and thus the electrons form a two-dimensional array of quantum dots with two different diameters.

In terms of predicted results, this proposal focused on "stripe formation" in the Cu-O planes at filling factor 1/8 in the array. Interestingly, the proposal predicted that stripes would also form in the quantum dot array, as fluctuations in the chemical potential or electrostatic voltage. We call this situation "interesting," because most AQS proposals do not have such direct geometrical correlation between the AQS device and the underlying material described by the

Hamiltonian. Manousakis proposed measuring this stripe formation in the quantum dot array by electric force microscopy. Manousakis also proposed pursuing transport measurements in the same array, but without detailed explication.

Somewhat later came a similar proposal,[8] again to simulate the Fermi–Hubbard model by generating a two-dimensional array of quantum dots, seen schematically in Fig. 33. One of the themes in this proposal was to reduce the disorder (a perennial issue in AQS with semiconducting quantum dots) by (i) using undoped rather than doped GaAs/AlGaAs heterostructure, and (ii) working at high density accumulated by the global gate. This latter feature of high density leads to less disorder because, in semiconductors, typically higher density leads to higher mobility and less scattering of carriers; this high density also leads to simulating only using upper conduction bands. Another theme in this proposal was that the lower (completely filled) conduction bands would provide screening, which would effectively reduce the on-site repulsion energy $U$ for the upper bands, and thus make it possible to reduce $U$ to the regime where $U \sim t$.

The main prediction in this proposal was to measure transport through the device and see a "V-shaped" feature in the phase space of chemical potential and amplitude of the array confining potential. This is a difficult measurement to achieve.

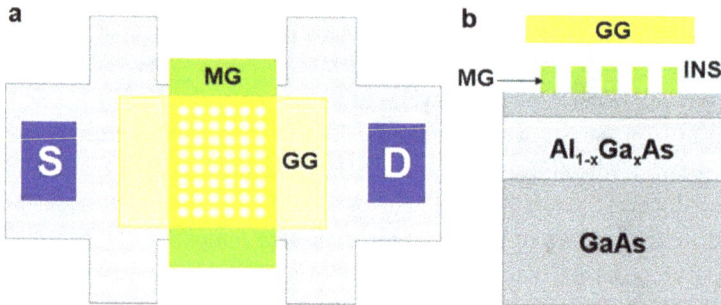

Fig. 33. Sketch of the proposed device, with MG comprising a depletion gate and (not shown) a continuous layer above as an accumulating global gate.

A recent interesting proposal, which focused on AQS using standard quantum dot array measurements, proposed to investigate the Heisenberg Hamiltonian using singlet-triplet states.[9] Unlike the Fermi–Hubbard model, the authors proposed to look explicitly at spin interactions and simulate the Heisenberg Hamiltonian. As seen in Fig. 34, they proposed a $2 \times N$ ($N$ of order 6 to 10) array of quantum dots, with read-out using a multiplexed RF gate reflectometry technique. They described in some detail how this technique would allow them to

(a)

(b)

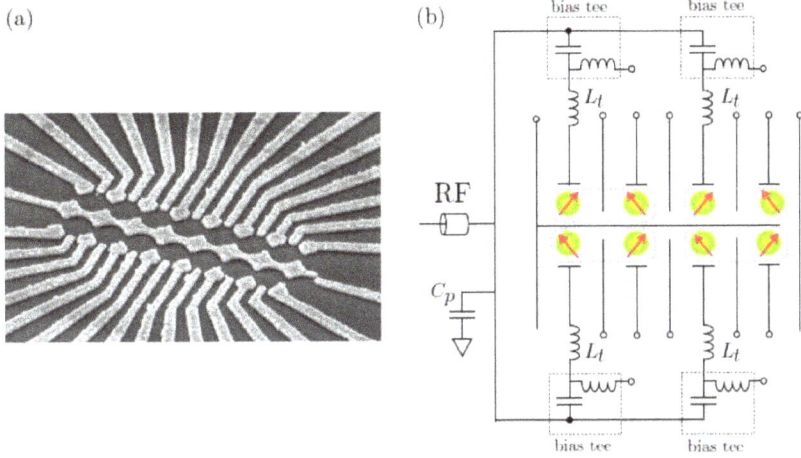

Fig. 34. A typical SEM micrograph of a 2 × N gate array, and the proposed measurement circuit to deduce the number of triplets in the array for a given set of simulator parameters. (From Ref. [9])

fully characterize the ground state in the simulator (the first goal of any AQS) without having to do full quantum tomography, which is quite challenging. They showed various predictions for measurable implications of the Heisenberg spin chain, including a quantum phase transition where the second-nearest-neighbor interaction becomes dominant, as well as deducing whether the thermal mixed state or the pure ground state is achieved. In general, these results would be deduced in the simulator by counting the number of triplets versus singlet states in pairs of quantum dots in the array.

For instance, Fig. 35 shows the probability distribution of the number of triplets in the array, and specifically observes a marked change in the distribution after one goes through the second-nearest-neighbor quantum phase transition. The authors also gave some thought to the two dominant noise sources — temperature and local magnetic field fluctuations from nuclear spins. In both cases, they predicted necessary values of temperature and/or relative variance of local magnetic field, which appear quite reasonable experimentally.

Finally, a very recent proposal takes quite a different tack to generating a quantum dot array, by using "acoustic tweezers"[10] *i.e.* using standing acoustic waves generated by counter propagating surface acoustic wave (SAW) transducers to trap electrons. The requirements to achieve such standing acoustic traps appear fairly challenging, both in terms of phonon emission rates and materials quality. In addition, they suggested an "exotic stability regime," which would obviate the need for low emission rate and high materials quality, but at the expense of very large SAW amplitudes. The suggested applications of this new

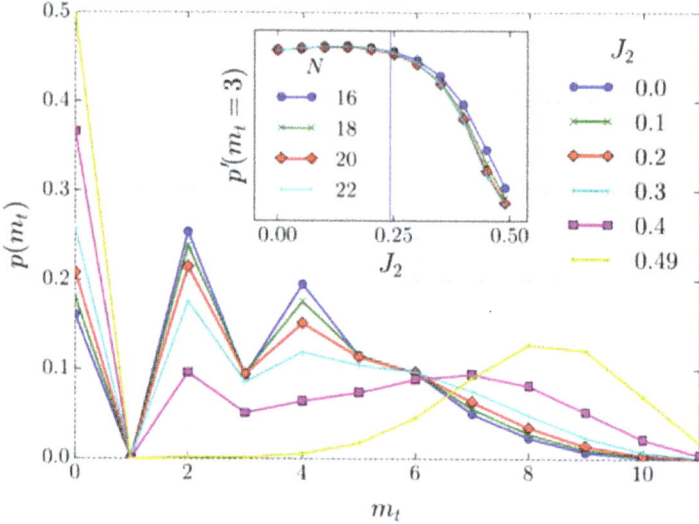

Fig. 35. Probability distribution of the number of triplets; there is a clear change in the distribution between the two sides of the second-nearest-neighbor quantum phase transition, located at $J_2 = 0.24$. (From Ref. [9])

method of generating quantum dots include: (i) AQS of the Fermi–Hubbard model, with read-out being variously optical detection, scanned probe detection, or DC transport. (ii) Going beyond AQS, the authors make the point that, as is well known in the field, moving quantum dots long distances with a fixed gate pattern can be challenging; instead, by varying the SAW parameters, one could move the quantum dot array around in a quite controlled fashion, leading to a number of new possibilities.

### Experimental results for AQS using semiconducting quantum dots

As discussed above, achieving AQS in semiconducting quantum dot arrays has not reached fruition, mostly because of the very important issue of disorder. On the other hand, multiple groups continue to work in this field, because of a very appealing parameter regime where the inter-particle interactions and spin interactions can be large compared to the thermal energy. Barthelemy[11] provides a useful "snapshot" of the experimental state-of-the-art before 2013.

There have been a number of studies of small "arrays" with two or three quantum dots, investigating simple Hubbard, Kondo, and Heisenberg Hamiltonians.[11] In these small-array devices, disorder can be tuned out by adjusting individual gates; typically, the AQS results achieved in these small-array devices are also achievable by theory or by simulation on conventional computers. In contrast, reducing the disorder by tuning individual gates was not

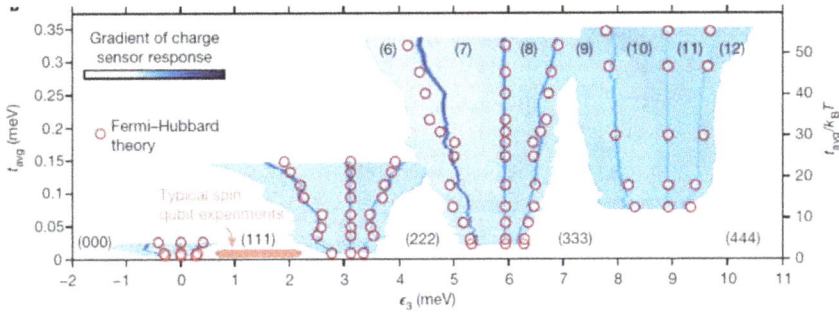

Fig. 36. Collective Coulomb blockade in a triple quantum dot. (From Ref. [12])

considered feasible in the large arrays, where new physics can be studied. Suggestions for decreasing the disorder included reducing dopants and background impurities (also mentioned above in Ref. [8]).

As a first step towards reducing the effects of disorder in a larger array, Hensgens et al.[12] showed automated tuning of a triple quantum dot, which they furthermore suggested could be scaled up to overcome the disorder problem. Transport measurements through the triple dot array as shown in Fig. 36 allowed deducing the filling of the array with up to nine electrons. Hensgens et al. characterized this as collective Coulomb blockade, attributed to the finite-size analogue of the interaction-driven Mott metal-to-insulator transition.

In an intriguing study, Wei et al.[13] added a new flavor to the mix, by applying a large magnetic field to a quantum dot array in GaAs. In this way, they forced

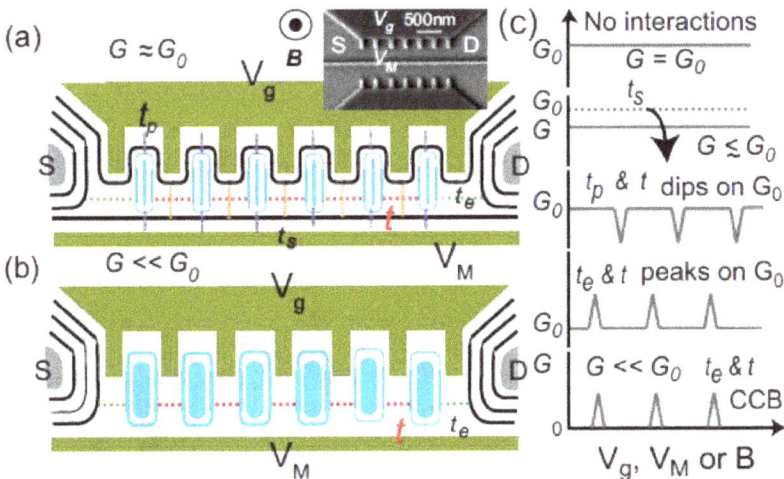

Fig. 37. Sketch of the proposed device, along with schematic measurements of conductance as a function of gate voltage, showing the transition from conducting to insulating as a function of either gate voltage or magnetic field. (From Ref. [13])

the electrons in the quantum dots into the integer quantum Hall regime, where the electrons exist only in edge states. The authors experimentally showed (i) coupling between the "bulk" quantum dot states and the edge states, (ii) controllable inter-dot coupling and dot occupation using a combination of both gate voltages and magnetic field, and (iii) concomitant conducting-to-insulating transition as a function of those control parameters (see Fig. 37). The authors went on to suggest that these edge-state-mediated multi-QDs could provide a good laboratory for exploring many-body interactions.

## Summary

To summarize, the attraction of pursuing AQS in semiconductor quantum dot arrays is driven by the exciting possibility of getting into the parameter regime $U > k_B T$. While there have been some preliminary experimental results, this field is impeded in large part by the difficulty of reducing disorder sufficiently to achieve the Hubbard model conditions. On the other hand, these devices do appear feasible for simulating the Anderson–Hubbard model, which explicitly considers disorder.

## References

1   F. Zwanenburg *et al.*, *Silicon quantum electronics*, Rev. Mod. Phys. **85**, 961 (2013).

2   J. Salfi, S. Roddaro, D. Ercolani, L. Sorba, I. Savelyev, M. Blumin, H. E. Ruda and F. Beltram, *Electronic properties of quantum dot systems realized in semiconductor nanowires*, Semiconductor Science and Technology, (2010).

3   B. Kaestner and V. Kashcheyevs, *Non-adiabatic quantized charge pumping with tunable-barrier quantum dots: a review of current progress*, Reports Prog. Phys. **78**, 103901, 1–27 (2015).

4   M. Veldhorst, C. H. Yang, J. C. C. Hwang, W. Huang, J. P. Dehollain, J. T. Muhonen, S. Simmons, A. Laucht, F. E. Hudson, K. M. Itoh, A. Morello, and A. S. Dzurak, *A two-qubit logic gate in silicon*, Nature, 1-5 (2015).

5   K. Nishiguchi, A. Fujiwara, Y. Ono, H. Inokawa, and Y. Takahashi, *Room-temperature-operating data processing circuit based on single-electron transfer and detection with metal-oxide-semiconductor field-effect transistor technology*, Appl. Phys. Lett. **88**, 18, 2004–2007 (2006).

6   Craig S. Lent, P. Douglas Tougaw, Wolfgang Porod and Gary H. Bernstein, *Quantum cellular automata*, Nanotechnology **4**, 49–57 (1993).

7   E. Manousakis, *A Quantum-Dot Array as Model for Copper-Oxide Superconductors: A Dedicated Quantum Simulator for the Many-Fermion Problem*, Journal of Low Temperature Physics **126**, Nos. 5/6 March (2002).

8   Tim Byrnes, Na Young Kim, Kenichiro Kusudo, and Yoshihisa Yamamoto, *Quantum simulation of Fermi-Hubbard models in semiconductor quantum-dot arrays*, Phys. Rev. B **78**, 075320 (2008).

9   Johnnie Gray, Abolfazl Bayat, Reuben K. Puddy, Charles G. Smith, and Sougato Bose, *Unravelling quantum dot array simulators via singlet-triplet measurements* Phys. Rev. B **94**, 195136 (2016).

10  M. J. A. Schuetz, J. Knörzer, G. Giedke, L. M. K. Vandersypen, M. D. Lukin, and J. I. Cirac, *Acoustic Traps and Lattices for Electrons in Semiconductors*, Phys. Rev. X **7**, 041019 (2017).

11  Pierre Barthelemy and Lieven M. K. Vandersypen, *Quantum Dot Systems: a versatile platform for quantum simulations*, Ann. Phys. (Berlin) **525**, 10–11, 808–826 (2013)

12  T. Hensgens, T. Fujita, L. Janssen, Xiao Li, C. J. Van Diepen, C. Reichl, W. Wegscheider, *Quantum simulation of a Fermi–Hubbard model using a semiconductor quantum dot array*, Nature **548**, 70 (2017).

13  Wen-Yao Wei, Tung-Sheng Lo, Chiu-Chun Tang, Chi-Te Liang, D. C. Ling, C. C. Chi, Chung-Yu Mou, Dennis M. Newns, Chang C. Tsuei, and Jeng-Chung Chen, *Edge-state-mediated collective charging effects in a gate-controlled quantum dot array*, Phys. Rev. B **95**, 155445 (2017).

## 12. Photons

Garnett Bryant

*National Institute of Standards and Technology*

There are several ways in which photons (photonics) can play an important role in the development of atom-based planar (1D or 2D) materials. The focus of the workshop was on the use of arrays of dopants or patches of dopants to perform quantum simulations, especially for the Fermi–Hubbard model. Such proposals typical envision some form of exchange coupling between spin qubits on adjacent donors to mediate the interactions and hopping in a Fermi–Hubbard model. However, such interactions can be very sensitive to inter-donor spacing with order of magnitude variations for changes in dopant position of one lattice site. Through spin-photon conversion, photons could be used, instead of exchange, to provide coupling between spin qubits. Recently, such a proposal has been analyzed in detail.[1] Such approaches in Si would require the use of deep donors such as $Se^+$ to operate in the optical regime and Si photonic cavities to enhance the spin-photon conversion. Strong light emission has been demonstrated.[1] The spin ground states can be long-lived at 4 K. Spin qubits can maintain their bulk decoherence properties because large magnetic fields and interface charge manipulation is not used.[1] Such a capability opens another way to operate arrays of dopants for simulations.

While the main interest of the workshop was on the use of atom-based arrays for quantum simulations, such arrays could also be used for a variety of compelling photonic applications. Several possibilities were discussed in the workshop to highlight what might be possible. For example, it has recently been shown theoretically that a single 2D layer of a transition metal dichalcogenide (TMD) can act as a perfect mirror.[2] Complete destructive interference between the field transmitted through the TMD layer and the incident field leaves only the field reflected by the TMD. Excitons excited in the single 2D TMD layer by the incident field are reemitted in the same mode as the incident field, as required by in-plane momentum conservation in a 2D translationally invariant layer, but out of phase with the incident field, leading to complete destructive interference even by a single layer. Recently, large excitonic reflectivity by monolayers of $MoSe_2$ has been observed experimentally.[3,4] These results suggest the possibility of constructing entire optical circuits using single layers of atomic based structures if dopants with optical transitions are used or if the arrays can support excitonic states.

Other photonic functionalities could be engineered with dopants in Si. Quantum memories for photonic qubits have been long sought.[5] Light can be stored in optical fibers, but losses become significant in meter to kilometer long fibers needed for reasonable storage times. Photonic states can be stored by slowing or stopping light with atomic vapors in centimeter cells. However, this approach is not amenable to chip-scale integration. However, rare-earth dopants in Si could provide micrometer-scale photonic memories when a cavity is used to provide enhanced coupling.[5,6] Here photons are captured and shared among multiple dopants, all at slightly different energies. The shared photonic state is revived by rephrasing on a time-scale defined by the energy splitting between dopants and is then reemitted. This revival defines the storage time. These quantum memories have been implemented with randomly placed rare-earth dopants. 2D arrays based on STM lithography could provide even greater functionality and open ways to integrate quantum memories with a diverse set of quantum technologies.

Atom-based clocks have become the standard for time-keeping, providing the precision needed for modern commerce and banking, precision tracking and mapping, and high-precision frequency standards. Clocks based on ultracold atoms take up full labs to run. Chip-scale atomic clocks can now be made but they require tiny vacuum chambers to isolate atoms from the environment. The power drain to run chip-scale atomic clocks remains significant. Atoms can also be isolated from the environment by placing them in a cage. This has been demonstrated for N in $C_{60}$.[7] This could also be done with dopants in Si provided the dopants have transitions insensitive to environmental effects. Such clock transitions have been demonstrated for Bi in Si.[8,9] Again, the use of 2D dopant arrays based on STM lithography might provide greater functionality and more ways to integrate atomic clocks with a diverse range of quantum technologies and for a wide range of applications that need on-chip clocks, for example, to do tracking in local environments such as a building or warehouse without relying on GPS, or to stabilize or identify signals in noisy environments.[7]

Single-photon emission from individual, localized defects in atomically thin semiconductor layers, such as $WSe_2$ and $WS_2$, have been demonstrated recently.[10] It is not clear whether the defects are actual defects or regions of local strain that can trap optical excitations. However other experiments have shown that regions of local strain can be engineered by draping the 2D semiconductor layer over an array of lithographically formed pillars on a substrate and that the pillar-induced strained regions can act as quantum emitters with single-photon emission.[11] Such arrays of quantum emitters in two-dimensional materials offer promising applications in quantum computing, quantum communication, quantum sensing and fundamental quantum science.[12] Atom-based arrays in Si could provide another way to achieve ordered arrays of quantum emitters in planar structures, allowing for atomically precise placement of the emitters and controlled integration with the other atom-based optical elements already mentioned, such as clocks, memories, and routing elements to create quantum optical circuits. All that is needed to complete such quantum optical circuits is a way to get light in and out. Metal nanoparticles (MNP) are now often used as nanoantennae because plasmonic excitations in MNPs can provide the intense fields in nanoscale volumes needed for efficient coupling to single quantum emitters. An open question is whether atom-based plasmonic structures could be engineered from 1D and 2D arrays of precisely placed dopants.

So far, we have discussed ways in which photons and photonics could play an important role in the development or application of atom-based planar (1D or 2D) materials. The focus of this workshop has been on the use of arrays of dopants or patches of dopants to perform quantum simulations, especially for the Fermi–Hubbard model. Typically, such simulations are done to determine ground state phase diagrams and to understand the competing effects of many-body interactions, single-particle hopping, and quantum fluctuations in simple models. Applications of dopant arrays to photonics require an understanding of many-body excited states and response. Dopant arrays could be used both to do these excited state simulations and to realize the atom-based photonic structures being simulated. Dopant arrays could be used both to simulate excited state dynamics, many-body entanglement, collision, and interference of many-body excitations and to implement atom-scale quantum and photonic applications.

We now present an example of how this could be realized.[13] Dopant based arrays can be modeled to determine if, when and how plasmonics might appear in such arrays. In this example, we focus on atom-scale systems that are metal-like and can support many-body states that are plasmonic. We model small metal-like atomic scale systems to find the full spectrum of excitations and identify the quantum plasmons. The simplest model for an atom-scale system is a one-dimensional chain of atoms (the dopants). To describe a metallic system, we consider a single band defined by a hopping $t$ between nearest neighbor, atomic sites (see Fig. 38 below). We assume that the band is half-filled with spinless electrons. Thus, there is one electron per two sites. We use an extended range Hubbard model:

$$H = \sum_{\langle i,j \rangle_{nn}} t(c_i^\dagger c_j + c_j^\dagger c_i) + \sum_i V_{nuc}(i)n_i + \sum_{\langle i,j \rangle} V_{ee}^D(i,j)n_i n_j$$

where the first sum extends over all nearest neighbor pairs and $c_i^\dagger$ ($c_i$) is the creation (destruction) operator for an electron at atom $i$, $n_i = c_i^\dagger c_i$ is the electron number at atom $i$, and the third sum extends over all pairs that are coupled through the extended range interaction. For spinless electrons, Pauli blockade ensures that

$$c_i^\dagger c_{i+1} \qquad V_{ee}^D \qquad V_{ee}^{exch}$$

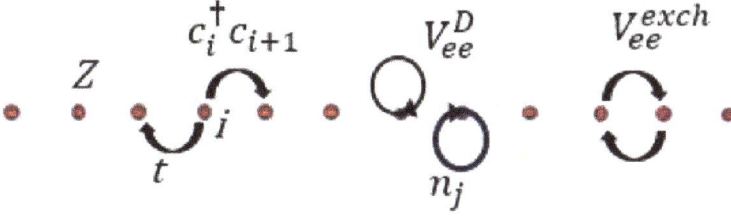

Fig. 38. Schematic of the 1D model.

there are never two electrons on the same site and there is no on-site interaction. Here all dopant atoms are identical, and the on-site energy is the same for each atom. We model the interactions with a long-range Coulomb interaction to best describe a plasmonic system. The extended-range Coulomb interaction energy $V_{nuc}$ between an electron at site $i$ and the atomic cores is

$$V_{nuc}(i) = -\sum_j \frac{\lambda_{nuc}Z}{|i-j| + \xi_{nuc}},$$

where the sum extends over all atoms $j$ and $Z$ is the nuclear charge. The direct extended-range, electron-electron Coulomb interaction energy $V_{ee}^D$ between electrons at sites $i$ and $j$ is

$$V_{ee}^D(i,j) = -\frac{\lambda_{ee}}{|i-j| + \xi_{ee}},$$

where $\lambda_{nuc}$ and $\lambda_{ee}$ are scale factors that include the length scale for the nearest-neighbor separation and any dielectric screening. $\xi_{nuc}$ and $\xi_{ee}$ are cutoffs that account for the spread of the electron orbital. Exchange effects can also be included by reducing the strength of the Coulomb interaction when two electrons are nearest neighbors:

$$V_{ee}^{exch}(i, i_{nn}) = -\lambda_{exch} V_{ee}^D(i, i_{nn}),$$

where $\lambda_{exch}$ is the scale of the nearest neighbor exchange and site $i_{nn}$ is a nearest neighbor to site $i$.

For a chain with $N_s$ sites and $m_e$ electrons, there are $\binom{N_s}{m_e}$ states. There are 924 states in a half-filled, 12-atom chain. Such chains are short enough that the entire Hamiltonian can be diagonalized, the full spectrum obtained, and all of the eigenstates determined. The number of states increases by roughly two orders of magnitude when four sites are added. A chain with 20 atoms has nearly two hundred thousand states, so several hundred lowest energy states can be found, but storage of more states becomes expensive. For a 28-site chain, there are more than forty million states and direct diagonalization of even a few states becomes intractable. Including additional effects, such as spin or coupling to a parallel chain of atoms to model two-dimensional effects that break the Pauli blockade of one-dimensional systems are straightforward. However, including either effect would reduce the chain length that could be studied by exact diagonalization by a half.

Results for longer chains are needed to understand the evolution of plasmonic states as the chain approaches nanoscale lengths; however, direct diagonalization is not possible. Results for longer chains, especially for ground states, can be found accurately by approximation methods.[14] However, quantum simulations could easily extend results to longer chains. The phase diagram for this model and the related $t$-$V$-$V'$ model with hopping $t$, nearest neighbor interaction $V$ and next nearest neighbor interaction $V'$ has been studied for nearly 30 years.[14-17] Only recently, it has been shown that four phases appear in the ground state: a metallic, Luttinger liquid phase, two charge density wave phases, and a bond order phase.[14] Quantum simulations could be done to answer important questions about many-body excitations and dynamics in larger systems, such as how many-body excitations entangle and mediate entanglement, not just between excitations but also between the electrons in the excitations, how the excitations collide and interfere, also probing to understand how this involves the individual electrons.

To show what is possible with simulations of excitations, we show exact results for the many-body excitation in short chains. As an example, we show how quantum plasmons can be identified and how they develop in short chains. The scales for direct and nuclear Coulomb interactions are chosen to be the same, $\lambda_{ee} = \lambda_{nuc}$, to ensure that the potentials cancel at large distances with the same charge. We obtain spectra as a function of the ratio $\lambda_{ee}/t$. Because the Coulomb scale decreases as $1/a$, where $a$ is the atom separation while the hopping decreases exponentially with $a$, small ratios correspond to closely spaced atoms with the largest hopping. Large ratios correspond to widely space atoms with weak hopping.

The excitation spectrum of a half-filled band of spinless electrons on a 12-atom chain[13] as a function of the nearest neighbor interaction strength $\lambda_{ee}$ is shown in Fig. 39. The metallic phase occurs for nearest neighbor interaction strength comparable to or less than the hopping. For larger interactions, the ground state develops a charge oscillation with twice the lattice period, the gap to the first excitation closes and the ground state becomes doubly degenerate for the two ways the charge density wave/Wigner crystal can lock into the lattice. This is the ground state phase transition previously identified,[14-17] but now revealed by the excited states.

A blowup of the low interaction regime is shown in Fig. 40. The excitation energies increase linearly from the single-particle limit as the interaction is turned on. In this regime of linear response, the onset of quantum plasmons can be

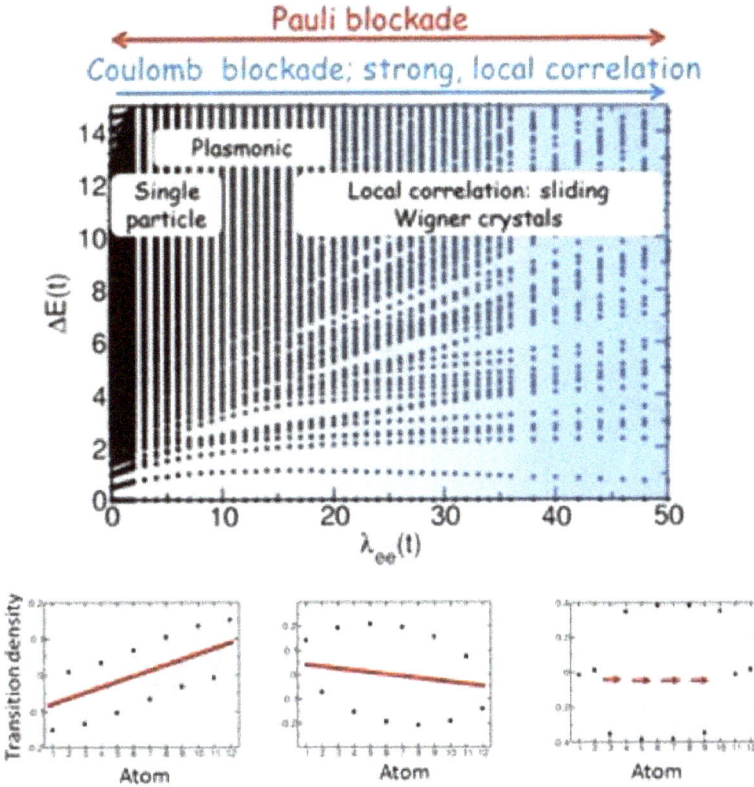

Fig. 39. Excitation spectrum of a 12-atom chain as a function of nearest neighbor interaction strength.[13] The oscillating charge density along the chain induced by an oscillating field along the chain is shown for the lowest-energy transition from the ground state for the regimes of single particle states (no interaction), the metallic regime (weak interactions and plasmons) and the regime of significant local correlation (strong interaction and sliding Wigner crystals).

identified. The first excitation corresponds to the lowest plasmon mode, $M = 1$, with a half wave of charge oscillation when driven by an oscillating field. The third excitation at almost twice the $M = 1$ plasmon energy can be identified as the excitation of two $M = 1$ plasmons. The sixth excitation at almost three plasmon energies is the excitation of three $M = 1$ plasmons. These three excitations correspond to the first three states of a bosonic ladder and are clear evidence of the bosonic statistics of these many-electron excitations. The second (fourth) excitation can be identified as the first excitation of the $M = 2$ ($M = 3$) mode with excitation slightly less than two (three) plasmon energies and a full wave (three half waves) of charge oscillation. This shows how quantum plasmons evolve out of the single-particle excitations in short chains, where excitations exhibit bosonic

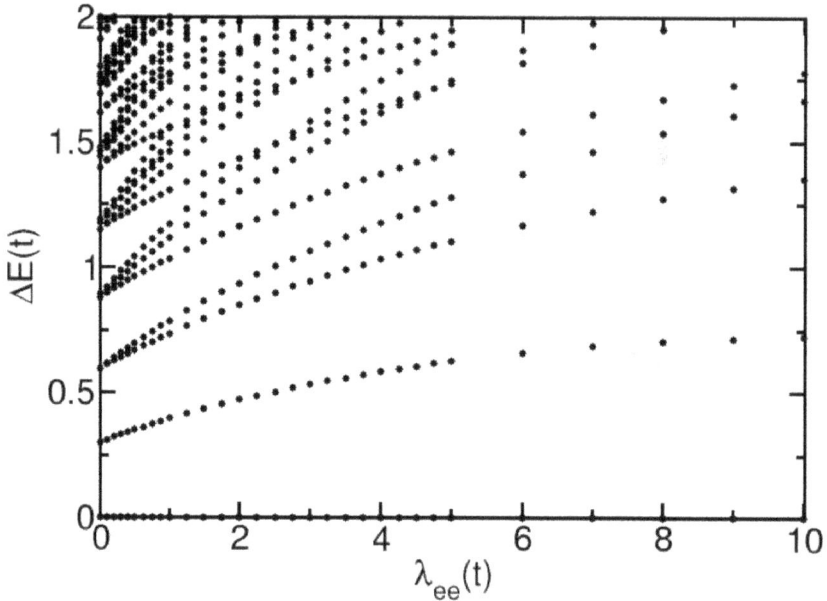

Fig. 40. Low energy excitations in a 20-atom chain in the metallic regime with weak interaction.[13]

character, and how this bosonic quantization breaks down for higher excitations. The many-body wave functions can be analyzed to understand electron entanglement in the plasmonic states.[13] Quantum simulations with dopant arrays are needed to show how this evolution continues toward the classical limit for larger systems and to address more complicated systems with spin or higher dimension.

The computational simulations presented here considered short, perfectly regular chains of identical atoms as a function of the interaction strength relative to the hopping energy. Quantum simulations with fabricated dopant arrays are needed to extend the results to longer chains, especially to explore the dynamics of many-body excitations. However, in such simulations the disorder in the dopant placement will be a critical issue. Recent results[18] suggest the robustness of simulations with dopant arrays to disorder in dopant placement. However, much still needs to be done to understand how sensitive the results will be to disorder, especially when excitations are simulated. Development of improved precision of dopant placement would help to eliminate the effects of disorder and, at the same time, provide a way to develop controlled tests of disorder. In computational simulations, it is easy to vary the interaction strength. In a physical system the interaction strength, relative to the hopping, will be determined by the spacing between dopants. Widely space dopants will be in the strong interaction regime because the hopping becomes small exponentially with separation. The opposite limit of weak interaction is realized with the dopants are spaced by about a lattice constant apart. Coupling can further be modified by controlling the alignment of the array to the crystal axis to exploit valley mixing. A range of new simulations with arrays could be done by using different types of dopant atoms and different arrangements or orderings of the dopants. Developing this capability would greatly extend the type of simulations that could be done and could make assessable simulations of exotic materials with complex unit cells.

## References

1  K. J. Morse, R. J. S. Abraham, A. DeAbreu, C. Bowness, T. Richards, H. Riemann, N. V. Abrosimov, P. Becker, H.-J. Pohl, M. L. W. Thewalt and S. Simmons, *A photonic platform for donor spin qubits in silicon*, Science Advances, *e1700930, (2017)*.

2  S. Zeytinoglu, C. Roth, S. Huber and A. Imamoglu, *Atomically thin semiconductors as nonlinear mirror*, Physical Review A **96**, 031801 (2017).

3  P. Back, S. Zeytinoglu, A. Ijazz, M. Kroner and A. Imamoglu, *Realization of an electrically tunable narrow-bandwidth atomically thin mirror using monolayer MoSe$_2$*, Physical Review Letters **120**, 037401 (2018).

4  G. Scuri, Y. Zhou, A. A. High, D. S. Wild, C. Shu, K. DeGreve, L. A. Jauregui, T. Taniguchi, K. Watanabe, P. Kim, M. D. Lukin and H. Park, *Large excitonic reflectivity of a monolayer MoSe$_2$ encapsulated in hexagonal boron nitride*, Physical Review Letters **120**, 037402 (2018).

5  E. Waks and E. A. Goldschmidt, *Storing light in a tiny box*, Science **357**, 1354 (2017).

6  T. Zhoung, J. M. Kinden, J. G. Bartholomew, J. Rochman, C. I. E. Miyazono, M. Bettinelli, E. Cavalli, V. Verma, S. W. Nam, F. Marsili, M. D. Shaw, A. D. Beyer and A. Faraon, *Nanophotonic rare-earth quantum memory with optically controlled retrieval*, Science **357**, 1392 (2017).

7  K. Porfyrakis and E. A. Laird, *Keeping perfect time with caged atoms*, IEEE Spectrum, **34,** (2017).

8  S. Rogge and M. J. Sellars, *Atomic clocks in the solid state*, Nature Nanotechnology **8**, 544 (2013).

9  G. Wolfowicz, A. M. Tyryshkin, R. E. George, H. Riemann, N. V. Abrosimov, P. Becker, H.-J. Pohl, M. L. W. Thewalt, S. A. Lyon and J. J. L. Morton, *Atomic clock transitions in silicon-based spin qubit,* Nature Nanotechnology **8**, 561 (2013).

10  C. Palacios-Berraquero, M. Barbone, D. M. Kara, X. Chen, I. Goykhman, D. Yoon, A. K. Ott, J. Beitner, K. Watanabe, T. Taniguchi, A. C. Ferrari and M. Atature, *Atomically thin quantum light-emitting diodes,* Nature Communications **7,** 12978 (2016).

11  C. Palacios-Berraquero, D. M. Kara, A. R.-P. Montblanch, M. Barbone, P. Latawiec, D. Yoon, A. K. Ott, M. Loncar, A. C. Ferrari and M. Atature,

144

*Large-scale quantum-emitter arrays in atomically thin semiconductors* Nature Communications **8,** 15093 (2017).

12    I. Aharonovich and M. Toth, *Quantum emitters in two dimensions*, Science **358,** 170 (2017).

13    G. W. Bryant, E. Townsend, T. Neuman, A. Debrecht and J. Aizpurua, (in preparation).

14    T. Mishra, J. Carrasquilla and M. Rigol, *Phase diagram of the half-filled one-dimensional t-V-V' model*, Physical Review B **84,** 115135 (2011).

15    V. J. Emery and N. C., *Critical properties of a spin-1/2 chain with competing interactions*, Physical Review Letters **60,** 631 (1988).

16    K. Hallberg, E. Gagliano and C. Balseiro, *Finite-size study of a spin-½ Heisenberg chain with competing interactions: Phase diagram and critical behavior*, Physical Review B **41,** 9474 (1990).

17    D. Poilblanc, S. Yunoki, S. Maekawa and E. Dagotto, *Insulator-metal transition in one dimension induced by long-range electronic interactions*, Physical Review B **56,** 1645 (1997).

18    A. Dusko, A. Delgado, A. Saraiva and B. Koiller, *Adequacy of Si:P chains as Fermi-Hubbard simulators*, Nature Partner Journals Quantum Information **4,** 1 (2018).

# Ancillary Appendix

## Workshop program

2D QMM Workshop Apr. 25,26, 2018, Gaithersburg, MD

| Wednesday 25th April | | | |
|---|---|---|---|
| **8 AM** | | **2 PM** | APM – Enrico Prati, IFN CRN, "Critical quantum chaos and room temperature effects in 1D arrays of P donors in silicon" |
| :15 | | :15 | |
| :30 | Registration and Continental Breakfast | :30 | |
| :45 | | :45 | APM – Jon Wyrick, NIST, "Fabrication of atomic-precision dopant arrays in Si using STM-based hydrogen lithography" |
| **9 AM** | | **3 PM** | |
| :15 | Welcome – Kent Rochford, ADLP, NIST | :15 | |
| :30 | NSF - Cooper and Pavlidis | :30 | Coffee Break |
| :45 | | :45 | |
| **10 AM** | Plenary 1– Shashank Misra, Sandia, "Designing quantum materials, atom by atom" | **4 PM** | |
| :15 | | :15 | Alternatives - Cheng Chin, Chicago: (experiment) cold atoms |
| :30 | | :30 | |
| :45 | Plenary 2 – Gabe Aeppli, PSI, "Engineering quantum many-body physics" | :45 | |
| **11 AM** | | **5 PM** | Alternatives – Alicia Kollar, Princeton: (Experiment) Superconducting TBD |
| :15 | | :15 | |
| :30 | | :30 | Alternatives – Sjaak van Diepen, TU Delft: (Experiment) Semiconductor TBD |
| :45 | Lunch | :45 | |
| **12 PM** | | **6 PM** | |
| :15 | | :15 | Shuttles back to hotel |
| :30 | | :30 | Social Event sponsored by ScientaOmicron |
| :45 | | :45 | |
| **1 PM** | Plenary 3 – Subir Sachdev, Harvard, "The disordered Hubbard model: from Si:P to the high temperature superconductors" | **7 PM** | |
| :15 | | :15 | |
| :30 | | :30 | |
| :45 | | :45 | |

2D QMM Workshop Apr. 25,26, 2018, Gaithersburg, MD

| | Thursday 26th April | | |
|---|---|---|---|
| **8 AM** | | **2 PM** | |
| :15 | Continental Breakfast | :15 | |
| :30 | | :30 | |
| :45 | Welcome (John Randall) | :45 | Breakout sessions (2:00 total) |
| **9 AM** | APM - Ingmar Swart, Univ. Utrecht (Experiment), TBD | **3 PM** | |
| :15 | | :15 | |
| :30 | | :30 | |
| :45 | Alternatives – Kaden Hazard, Rice, "Ultracold matter for quantum simulations: achievements, possibilities, and challenges" | :45 | |
| **10 AM** | | **4 PM** | |
| :15 | | :15 | Reports by breakout leaders |
| :30 | Alternatives – Norbert Linke, U. Maryland, "Quantum simulation with trapped atomic ions" | :30 | |
| :45 | | :45 | |
| **11 AM** | | **5 PM** | |
| :15 | Alternatives – Garnett Bryant, NIST, "Atom-based devices for photonics, quantum plasmonics and many-body physics" | :15 | Final session: Discussion and final comments |
| :30 | | :30 | |
| :45 | | :45 | |
| **12 PM** | | **6 PM** | |
| :15 | Lunch | :15 | End |
| :30 | | :30 | |
| :45 | | :45 | |
| **1 PM** | Perspective – Philip Phillips, UIUC, "Metamaterials from Correlated Disorder and non-local Electromagnetism" | **7 PM** | |
| :15 | | :15 | |
| :30 | | :30 | |
| :45 | Instructions for breakout sessions | :45 | |